新一代 信息技术
"十三五"系列规划教材

Android
移动开发
基础教程 慕课版

◆刘刚 主编　　◆高伟南 副主编

人民邮电出版社

北京

图书在版编目（CIP）数据

Android移动开发基础教程：慕课版 / 刘刚主编. -- 北京：人民邮电出版社，2019.7
新一代信息技术"十三五"系列规划教材
ISBN 978-7-115-50879-9

Ⅰ. ①A… Ⅱ. ①刘… Ⅲ. ①移动终端－应用程序－程序设计－教材 Ⅳ. ①TN929.53

中国版本图书馆CIP数据核字（2019）第034925号

内 容 提 要

本书详细讲解了 Android 软件开发的基本方法和常用技能。全书分为 9 章，内容包括 Android 入门、Android 界面开发、Activity、Intent 和 BroadCastReceiver、数据存储、ContentProvider、Service、高级编程及综合实战，通过大量实例展示相关技术与技巧，最后通过完整项目的开发实现过程来提高读者的综合开发水平。

本书内容结构清晰，基本概念和机制讲解通俗易懂，案例丰富实用，适合作为高等院校、高职高专计算机及相关专业移动应用开发课程的教材，也适合 Android 爱好者自学和开发人员参考。

◆ 主　　编　刘　刚
　 副 主 编　高伟南
　 责任编辑　桑　珊
　 责任印制　马振武

◆ 人民邮电出版社出版发行　北京市丰台区成寿寺路 11 号
　 邮编 100164　电子邮件 315@ptpress.com.cn
　 网址 http://www.ptpress.com.cn
　 固安县铭成印刷有限公司印刷

◆ 开本：787×1092　1/16
　 印张：13.5　　　　　2019 年 7 月第 1 版
　 字数：326 千字　　　2025 年 1 月河北第 9 次印刷

定价：45.00 元

读者服务热线：(010)81055256　印装质量热线：(010)81055316
反盗版热线：(010)81055315
广告经营许可证：京东市监广登字20170147号

前言
Foreword

本书全面贯彻党的二十大精神，以社会主义核心价值观为引领，传承中华优秀传统文化，坚定文化自信，使内容更好体现时代性、把握规律性、富于创造性。

Android 简介

Android 系统是目前世界上市场占有率最高的移动操作系统，已经占据了全球智能手机操作系统 70%以上的份额。Android 从面世以来，已经发布了二十多个版本，众多开发者也为 Android 制作了丰富的应用程序，手机厂商、开发者、用户共同建立了一个完整的生态系统，推进了 Android 的蓬勃发展。

本书特色

本书细致地讲解了 Android 的基础知识，从 Android 入门到 Android 必备基础知识，再到高级编程，最后通过综合实例进行实践，内容全面，循序渐进，并针对重点知识点设计了案例训练，在重点章节之后设置小练习，帮助读者边学边练，掌握重点难点。全书最后精选视频播放器案例，带领读者实现典型应用功能，体验 Android 在实际工作中的应用，便于与实际开发工作接轨。

本书配套资源

本书配套慕课由一线程序员小刚老师（刘刚）详细讲解。登录人邮学院网站（www.rymooc.com）或扫描封面上的二维码，使用手机号注册，在首页右上角单击"学习卡"选项，输入封底刮刮卡中的激活码，即可在线观看全书慕课视频。扫描书中的二维码也可以直接使用手机观看视频。

本书全部案例源代码、素材、最终文件、电子教案、教学大纲、教案、实训可登录人邮教育社区（www.ryjiaoyu.com.cn）免费下载使用。

小刚老师（刘刚）简介

- 一线项目研发、设计、管理工程师，高级项目管理师、项目监理师，负责纪检监察廉政监督监管平台、国家邮政局项目、政务大数据等多个国家级项目的设计与开发。
- 极客学院、北风网金牌讲师。
- 畅销书《微信小程序开发图解案例教程（附精讲视频）》《小程序实战视频课：微信小程序开发全案精讲》《Axure RP8 原型设计图解微课视频教程 （Web+App）》作者。

编者
2022 年 12 月

目录 Contents

第1章 Android 入门 1
1.1 Android 的起源和发展 2
1.1.1 Android 的起源 2
1.1.2 Android 平台架构 2
1.2 Android 开发环境搭建 3
1.2.1 Android Studio 的安装 3
1.2.2 Gradle 文件介绍和常用设置 5
1.2.3 模拟器的创建和使用 7
1.3 本章小结 9

第2章 Android 界面开发 10
2.1 视图组件与视图容器 11
2.2 常用布局 12
2.2.1 线性布局 12
　　案例 2.1　使用线性布局 13
　　案例 2.2　使用嵌套的线性布局 15
2.2.2 相对布局 17
　　案例 2.3　使用相对布局 17
2.2.3 列表视图 20
　　案例 2.4　通过数组资源文件填充数据 21
　　案例 2.5　通过 Adapter 填充数据 22
　　案例 2.6　通过自定义 Adapter 填充数据，显示学生考试信息 22
2.2.4 网格视图 26
　　案例 2.7　以网格的形式排列显示 1～9 个数字 27
2.3 常用控件 30
2.3.1 文本框和编辑框 30
　　案例 2.8　显示不同颜色、大小和不同位置的文字 31
　　案例 2.9　文字超长时的处理 32
　　案例 2.10　将指定格式的文本转化为可单击的链接 34
2.3.2 按钮 36
　　案例 2.11　切换"Hello"和"World"的显示 36
2.3.3 单选按钮和复选框 38
　　案例 2.12　选择性别与爱好 38
2.3.4 图片控件 42
　　案例 2.13　图片尺寸大于 ImageView 控件尺寸的大小 43
2.3.5 进度条和拖动条 47
　　案例 2.14　使用进度条 47
　　案例 2.15　使用拖动条 49
2.4 对话框 51
2.4.1 简单对话框 51
　　案例 2.16　使用简单对话框 52
2.4.2 列表对话框 53
　　案例 2.17　使用列表对话框选择语言 53
2.4.3 自定义对话框 55
　　案例 2.18　使用自定义对话框制作登录页面 55
2.5 菜单 57
2.5.1 选项菜单 57
　　案例 2.19　制作"添加""删除""查询"和"退出"选项菜单 59
2.5.2 上下文菜单 61
　　案例 2.20　制作"添加""删除""查询"和"退出"上下文菜单 61
2.6 常用资源类型 63
2.6.1 资源的类型和使用 63

2.6.2 字符串、颜色、尺寸 65
    🔗 案例 2.21 字符串、颜色、尺寸的具体定义和使用 66
2.6.3 Drawable 67
    🔗 案例 2.22 使用图片资源 67
    🔗 案例 2.23 使用 State List 制作按钮按下变色效果 69
    🔗 案例 2.24 使用 Shape Drawable 制作圆角矩形的编辑框 70
2.6.4 Style 72
    🔗 案例 2.25 使用 Style 统一设置文字的大小和颜色 72
2.6.5 国际化 73
    🔗 案例 2.26 制作同样的按钮在不同的语言环境下的显示效果 74
2.7 事件处理和消息传递 75
  2.7.1 基于监听的事件处理 75
  2.7.2 基于回调的事件处理 75
    🔗 案例 2.27 基于回调事件的处理 76
  2.7.3 Handler 消息传递 76
    🔗 案例 2.28 基于回调事件的处理 77
2.8 本章小结 78

第 3 章 Activity 79

3.1 Activity 的使用 80
3.2 Activity 之间的跳转 80
    🔗 案例 3.1 用 startActivity 方法实现跳转 81
    🔗 案例 3.2 用 startActivityForResult 方法实现登录效果 83
3.3 Activity 的生命周期 85
3.4 Activity 的启动模式 89
3.5 本章小结 90
3.6 小练习 90

第 4 章 Intent 和 BroadCastReceiver 97

4.1 Intent 和 intent-filter 配置 98

    🔗 案例 4.1 通过 action 启动 activity 98
4.2 BroadCastReceiver 101
  4.2.1 广播机制介绍 101
  4.2.2 静态注册 101
    🔗 案例 4.2 BroadCastReceiver 的使用 102
  4.2.3 动态注册 103
    🔗 案例 4.3 动态注册广播 104
  4.2.4 系统广播介绍 105
    🔗 案例 4.4 通过接收系统广播提示用户充电 105
4.3 本章小结 106

第 5 章 数据存储 107

5.1 SharedPreferences 108
  5.1.1 SharedPreferences 与 Editor 简介 108
  5.1.2 SharedPreferences 存储的位置和格式 108
    🔗 案例 5.1 使用 SharedPreferences 存储数据 108
5.2 File 存储 110
  5.2.1 读写内部存储 111
    🔗 案例 5.2 使用内部存储 111
  5.2.2 读写外部存储 113
5.3 SQLite 数据库 114
  5.3.1 SQLiteDatabase 简介 114
  5.3.2 创建数据库和表 115
    🔗 案例 5.3 创建表，存储学生考试成绩 115
  5.3.3 操作 SQLite 常用类 116
    🔗 案例 5.4 访问 SQLite 数据库，修改学生成绩表 116
  5.3.4 事务 121
5.4 本章小结 122
5.5 小练习 123

第 6 章 ContentProvider 129

6.1 ContentProvider 和 URI 简介 130
6.2 创建 ContentProvider 131
    🔗 案例 6.1 创建 ContentProvider，对外提供学生信息 131

6.3 使用 ContentResovler 操作数据 136
  案例 6.2 使用 ContentResovler 添加、查询联系人 136
6.4 本章小结 138
6.5 小练习 139

第 7 章 Service 143

7.1 Service 简介 144
 7.1.1 创建、配置 Service 144
 7.1.2 启动和停止 Service 145
  案例 7.1 启动和停止 Service 145
 7.1.3 绑定 Service 147
  案例 7.2 绑定和解绑 Service 148
7.2 Service 的生命周期 152
7.3 跨进程调用 Service 152
 7.3.1 创建 Service 和 AIDL 接口 152
 7.3.2 跨进程绑定 Service 154
7.4 本章小结 157
7.5 小练习 157

第 8 章 高级编程 161

8.1 网络编程 162
 8.1.1 TCP、UDP 协议基础 162
 8.1.2 Socket 通信 162
  案例 8.1 实现网络通信 163
 8.1.3 下载网络资源 166
  案例 8.2 下载网络图片 166
8.2 图形图像和动画 167
 8.2.1 Bitmap 和 BitmapFactory 167
 8.2.2 Android 绘图基础 168
  案例 8.3 使用线性布局 169
 8.2.3 补间动画 171
  案例 8.4 使用补间动画 171
 8.2.4 属性动画 173
  案例 8.5 使用属性动画 173
8.3 多媒体应用开发 175
 8.3.1 MediaPlayer 类介绍 175
 8.3.2 使用 MediaPlayer 和 SurfaceView 播放视频 176
  案例 8.6 使用 MediaPlayer 和 SurfaceView 播放视频 176
8.4 线程开发 180
 8.4.1 AsyncTask 及其使用 180
 8.4.2 ThreadPoolExecutor 介绍 181
8.5 Fragment 182
 8.5.1 Fragment 的创建 183
  案例 8.7 单击底部按钮，上面内容区域动态改变 184
 8.5.2 Fragment 的生命周期 189
  案例 8.8 通过日志打印看 Fragment 生命周期方法的回调顺序 190
8.6 RecyclerView 194
 8.6.1 RecyclerView 相关类 194
 8.6.2 RecyclerView 的使用 194
  案例 8.9 使用 RecyclerView 195
8.7 本章小结 198

第 9 章 综合实战 199

9.1 视频播放器 200
 9.1.1 界面布局 200
 9.1.2 初始化 202
 9.1.3 播控和进度控制 206
 9.1.4 横屏设置 209
9.2 本章小结 210

第1章

Android入门

■ 近年来，随着科技的发展，移动智能设备变得越来越普及。Android 是应用非常广泛的一种操作系统，可以运行在智能手机、平板电脑或其他移动设备上。目前，全世界有数十亿的智能设备使用的是 Android 系统，其使用范围遍及 190 多个国家。Android 以其开源性和简洁性而广受开发者的喜爱，通过 Android 可以开发出各种各样的应用程序、游戏等。本章我们首先来简要介绍一下 Android 的起源和发展，接着会重点介绍如何搭建 Android 的开发环境。

1.1 Android 的起源和发展

1.1.1 Android 的起源

Android 的首创者是 Andy Rubin，谷歌公司在 2005 年 8 月收购了 Andy Rubin 的创业公司，然后在 2007 年 11 月对外展示了 Android 操作系统，并宣布与多家制造商共同研发和改良 Android 系统。2008 年 9 月，谷歌正式发布了 Android 1.0 系统，内置谷歌移动服务，支持网络浏览、多任务处理、Wi-Fi、蓝牙和即时通信等功能。随后在 2009 年 4 月，谷歌发布了 Android 1.5 版本，并从这之后，每个版本开始以甜品的名字命名。例如，2017 年 8 月发布的 Android 8.0 版本名称为 Oreo（奥利奥）；2018 年将发布的 Android 9.0 初步代号定为 Pistachio Ice Cream（开心果冰淇淋）。

使用 Android 系统的手机现在已经成为市场占有率最高的手机，许多知名手机厂商如三星、HTC、小米、魅族、华为、中兴等，其移动设备的开发均是基于 Android 系统。

1.1.2 Android 平台架构

Android 系统的底层基础是 Linux 内核，Android 体系结构主要分为 4 层：应用程序层、Java API 框架层、硬件抽象层、Linux 内核层，具体如图 1.1 所示。

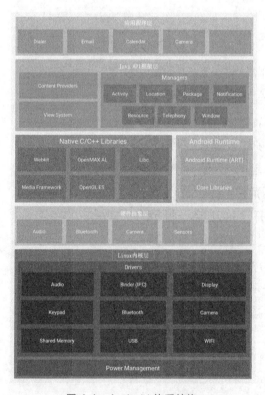

图 1.1 Android 体系结构

（1）应用程序层：Android 系统中的应用，包括电子邮件、日历、短信、照相机等，本书我们介绍的也是应用程序层的开发。

（2）Java API 框架层：Android 系统给开发者提供的开发接口，使用 Java 语言编写。通过这些接口，开发者可以构建自己的应用程序。

（3）硬件抽象层：向 Java API 框架层提供设备硬件功能。例如，当 API 需要访问照相机或蓝牙等硬件设备时，硬件抽象层为硬件组件加载对应的库模块。

（4）Linux 内核层：Android 系统基于 Linux 内核实现内存管理、线程调度、硬件资源分配等操作系统级别的功能。

1.2 Android 开发环境搭建

Android 开发最开始使用的编译器是 Eclipse，可以在 Eclipse 中集成 Android 的 SDK 和 ADT 完成开发环境的搭建。但是在 2013 年，谷歌公司推出了新的 Android 开发环境 Android Studio，并在 2015 年发布了正式版本。相比于 Eclipse，Android Studio 的代码提示和搜索功能更加智能，可支持设备预览的 UI 编辑器也使得开发者的工作变得更加高效。另外，Android Studio 还内置终端，可以直接输入命令；集成各种插件，如代码管理工具 git。我们需要不断使用开发工具才会对其进行熟练使用，本节我们将介绍一些 Android Studio 的基本内容。

1.2.1 Android Studio 的安装

首先下载 Android Studio 的安装程序，下载完成之后双击 exe 文件，即可开始安装 Android Studio，如图 1.2 所示。

图 1.2　Android Studio 开始安装界面

一直单击"Next"按钮，会出现选择安装目录页面，如图1.3所示。

图1.3　Android Studio 选择安装目录

选择好安装目录之后，一直单击"Next"按钮即可成功安装。Android Studio 第一次启动界面如图1.4所示，其中"Start a new Android Studio project"可以新建一个 Android 项目；"Open an existing Android Studio project"可以打开一个已有的 Android Studio 项目；"Check out project from Version Control"可以从一个代码管理工具上下载项目并打开；"Import project"可以打开一个符合 Android Studio 配置的项目；"Import an Android code sample"可以打开一个 Android Studio 示例代码。

图1.4　Android Studio 启动界面

1.2.2 Gradle 文件介绍和常用设置

在 Android Studio 中新建或打开一个 Android 项目后，工作台界面如图 1.5 所示。左侧是项目的组织结构，下拉可以选择不同的组织方式；上方是一些常用的工具栏和菜单栏；右侧是代码的编辑区；下方是一些常见的视图，如可以输入命令的终端、可以查看运行情况的监控台等。

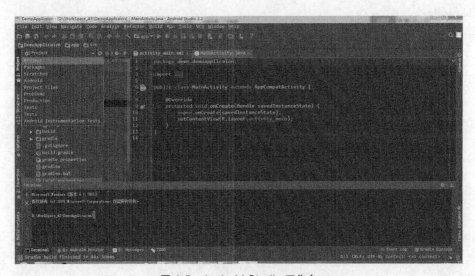

图 1.5　Android Studio 工作台

现在我们来具体看一下 Android Studio 的代码组织结构，在左侧的下拉列表中选择 "Project"，目录结构如图 1.6 所示。

图 1.6　Android Studio 的目录结构

Android Studio 使用 Gradle 工具对 Android 项目进行编译和构建，/gradle 目录中就存放了构建工具的 jar 包和 wrapper 等。/app 文件夹存放一个具体的项目，其中 /app/build 存放编译生成的 APK，/app/build/src 下面就是对应的 Java 源文件和 res 资源文件。/app 目录下还有一个很重要的文件 build.gradle，用于对项目做一些配置，大部分配置使用默认生成的即可，有些则要根据需要做一些更改。下面我们通过一个 build.gradle 的文件实例说明一下常见的配置。

build.gradle 文件：

```
apply plugin: 'com.android.application'   //Android应用插件，使用默认的即可

android {
    compileSdkVersion 24     //编译该程序时想要使用的API版本
    buildToolsVersion "25.0.2"    //构建工具的版本号
    defaultConfig {
        applicationId "demo.demoapplicaion"    //应用程序的包名
        minSdkVersion 15    //运行设备需要的Android系统最小版本
        targetSdkVersion 24    //目标设备的Android系统版本
        versionCode 1      //版本号
        versionName "1.0"    //版本号
        testInstrumentationRunner "android.support.test.runner.AndroidJUnitRunner"
    }
    buildTypes {   //定义如何构建APP
        release {
            minifyEnabled false
            proguardFiles getDefaultProguardFile('proguard-android.txt'), 'proguard-rules.pro'
        }
    }
}

dependencies {   //配置依赖包
    compile fileTree(dir: 'libs', include: ['*.jar'])
    androidTestCompile('com.android.support.test.espresso:espresso-core:2.2.2', {
        exclude group: 'com.android.support', module: 'support-annotations'
    })
    compile 'com.android.support:appcompat-v7:24.2.1'
    testCompile 'junit:junit:4.12'
}
```

除了版本号之外，其中经常会修改的配置为 dependencies 中的内容，有过 Java 开发经验的读者可

能比较清楚，如果需要在项目中使用额外的 jar 包，需要手动将需要的 jar 包下载下来，然后添加到项目中。而在 Android Studio 中，不需要手动下载，只需要在 build.gradle 文件的 dependencies 中进行配置即可，例如在上述 build.gradle 文件中添加了 compile 'com.android.support:appcompat-v7:24.2.1'，Gradle 工具会自动从远程仓库中引用 android-support-appcompat-v7.jar 的内容。

1.2.3 模拟器的创建和使用

在开发 Android 应用的过程中，经常需要对程序进行测试。一般可以在 Android 手机上进行调试，但是对于没有 Android 手机或者 Android 手机系统不符合要求的开发者来说，还可以使用模拟器进行调试。模拟器的创建也比较简单，首先从 Android Studio 的工具栏中找到 AVD Manager 图标，AVD Manager 图标如图 1.7 所示。

图 1.7 AVD Manager 图标

单击"AVD Manager"后开始创建模拟器，首先是选择模拟器的硬件信息，如图 1.8 所示，左侧可以选择 TV 设备、可穿戴设备、手机设备、平板设备，中间可以选择设备的尺寸、分辨率和密度，右侧是设备的实时预览。

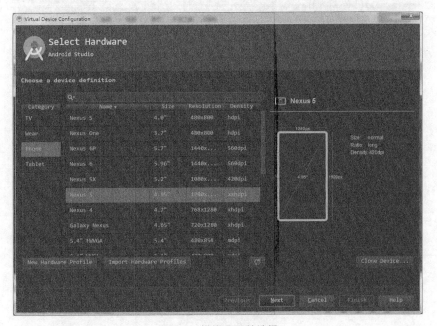

图 1.8 模拟器硬件选择

单击"Next"按钮可以选择设备的 Android 系统，如图 1.9 所示，可以选择系统版本和处理器的类型。

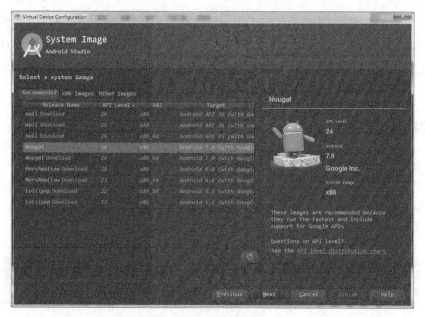

图 1.9　模拟器系统选择

再次单击"Next"按钮会进行模拟器信息的最终确认，如图 1.10 所示，单击"Finish"按钮可以完成模拟器的创建。

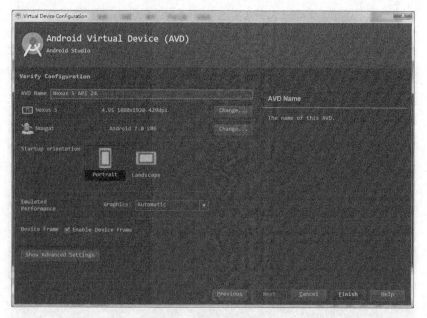

图 1.10　模拟器创建确认

模拟器创建完成之后，运行项目就会弹出模拟器选择界面，可选列表中会有之前创建的模拟器，如图 1.11 所示，选择一个模拟器并单击"OK"按钮，开发的 Android 项目会被安装到模拟器中并运行。

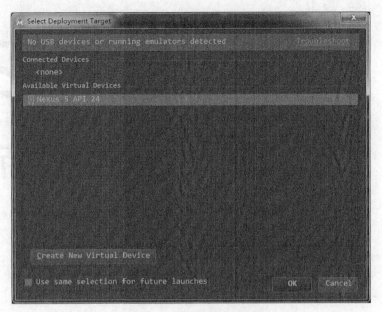

图 1.11　模拟器选择界面

1.3　本章小结

本章我们简要介绍了 Android 的背景知识，包括 Android 的起源和发展，并重点介绍了 Android 开发环境的搭建，包括项目的创建、目录的结构及常见文件的介绍。对于 Android Studio 等开发工具的熟练使用是开发 Android 应用必备的基础条件。最后介绍了模拟器的启动和使用，其可以用来调试应用程序。

第2章

Android界面开发

■ 在移动应用中,界面是给用户的第一印象。如果开发的 APP 界面粗糙、交互性差,即使它功能再强大,也很少有人愿意去使用。可以说,界面的美观与友好是吸引用户的先决条件。Android 为开发者提供了大量的控件,通过对这些控件进行组合,可以开发出各种各样的界面。本章我们就主要介绍 Android 开发中一些常用的控件。

2.1 视图组件与视图容器

在学习一些具体的控件之前，我们首先来了解几个基本的概念：控件、View（视图）和 ViewGroup（视图容器），然后再来了解开发用户界面的方式。控件是组成 Android 界面最基本的元素，每一个按钮或文本框都是控件。View 是所有控件的基类。ViewGroup 可以控制子控件的布局和显示。下面我们以图 2.1 为例，分别介绍这几个概念。

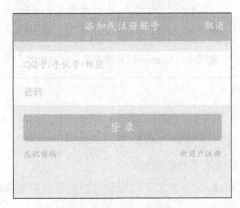

图 2.1 手机 QQ 登录页面

1. 控件

图 2.1 所示的是手机 QQ 的一个登录界面，界面中包含账号输入框、密码输入框、登录按钮等，这些基本元素称为控件或者组件，它们组合在一起形成了 Android 的用户界面。

2. View

View 是所有控件的基类，Android 界面中显示的所有控件都继承自 View，所以它是 Android 界面开发的基础。View 不仅包括控件的绘制，还包括一系列的事件处理，使得用户可以与界面进行交互。例如可以给图 2.1 所示的"登录"按钮设置监听事件，当用户单击按钮时，后台去执行登录操作。

在创建一个 View 的时候，开发者通常需要执行以下一些常见的操作。

（1）设置属性：View 的大小、位置、颜色等信息。

（2）设置焦点：View 是否需要获得焦点，在不同的操作下焦点如何移动。

（3）事件监听：有些时候，开发者需要对 View 设置一些监听事件，例如当 View 获得焦点或者被单击时可以做一些自定义的操作。

（4）可见性：开发者可以动态地控制 View 的显示或者隐藏。

3. ViewGroup

在 Android 中，管理控件的大小和控件之间的排列顺序称为布局管理，ViewGroup 可以实现布局管理。许多刚接触 Android 的人都不是太了解 ViewGroup 和 View 之间的关系，其实，ViewGroup 是 View 的一个重要子类。ViewGroup 可以理解为视图容器，开发者可以向其中添加一些基本的控件。例如在图 2.1 所示界面中，账号和密码两个输入框如何保持对齐，"登录"按钮如何居中显示，这些都可以通过将控件放入 ViewGroup 中来管理。ViewGroup 是一个抽象类，在开发过程中，通过实现它的子类来排列控件的显示位置。

Android 中所有的界面都建立在 View 和 ViewGroup 的基础上，ViewGroup 除了装载普通控件，还可以再次包含 ViewGroup 组件。图 2.2 是一个 Android 界面的层次结构图。

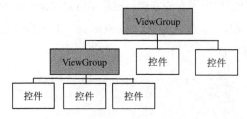

图 2.2　Android UI 界面的层次结构图

4．开发用户界面的方式

Android 支持使用以下两种方式定义用户界面。

（1）通过 Java 代码定义控件并设置控件的属性。

（2）通过 XML 文件控制控件的布局和属性。

通过 XML 布局文件控制 Android 的界面，可以使界面的设计更加简单清晰，具有更低的耦合性。而且这种方式可以将视图的逻辑从 Java 代码中抽离出来，更加符合 MVC（一种用模型—视图—控制器设计创建 Web 应用程序的模式）的设计原则。Android 也推荐使用 XML 的方式设计界面。

 耦合性是指模块与模块之间的紧密程度，一般都希望代码具有较低的耦合性，模块与模块之间的关系较为松散，就像汽车的各个零件一样，互相独立可拆卸，这样对一个模块的修改不会对另外一个模块产生较大的影响。

本节我们主要介绍了 View 和 ViewGroup 的基本概念，后续将结合示例介绍布局和控件的具体应用。

2.2　常用布局

在 Android 界面开发中，控件的布局非常重要，布局可以用来管理控件的分布和大小。不同的布局管理可以产生不同的布局效果，开发者需要根据不同的应用场景选择合适的布局管理，本节我们将介绍一些常用的布局方式。

2.2.1　线性布局

线性布局通过 LinearLayout 类来实现，LinearLayout 是前述 ViewGroup 的子类，是一个视图容器，可以向其中添加不同的控件。LinearLayout 将控件一个挨着一个排列起来，排列的顺序有横向排列和纵向排列。例如图 2.1 所示的手机 QQ 的登录页面中，账号输入框和密码输入框顺序排列，就可以通过将这两个控件放入 LinearLayout 中来实现。

第 2 章 Android 界面开发

在实际的开发过程中，会对 Android 原生的一些控件进行处理，使得界面更美观，如图 2.1 所示的 QQ 登录界面，文本框显示出来的效果和 Android 原生的控件有所不同。这里我们为了重点介绍线性布局，不对基本控件进行扩展，后续会在 2.6 小节介绍如何使用资源文件优化界面效果。

案例 2.1　使用线性布局

图 2.3 所示的是一个简化的登录界面的一部分，有一个文本框会提示用户输入用户名，后面紧跟着一个编辑框供用户输入具体的内容。

图 2.3　线性布局的示例——水平方向

对于这样的显示，可以使用横向线性布局来实现。

- **案例代码**

Activity 类：

```
public class MainActivity extends Activity
{
    protected void onCreate(Bundle savedInstanceState)
    {
        super.onCreate(savedInstanceState);
        setContentView(R.layout.act_layout_horizontal);    //加载布局文件
    }
}
```

XML 布局文件（act_layout_horizontal.xml）：

```
<?xml version="1.0" encoding="utf-8"?>
<LinearLayout xmlns:android="http://schemas.android.com/apk/res/android"
    android:layout_width="match_parent"
```

```
        android:layout_height="match_parent"
    android:orientation="horizontal" >      //横向布局
    <!-- 添加一个文本框控件显示"用户名" >
        <TextView
            android:id="@+id/txt_username"  //控件ID，名称可以自定义
            android:layout_width="wrap_content"   //控件宽度
            android:layout_height="wrap_content"  //控件高度
            android:gravity="left"   //控件内容的对齐方式
            android:textSize="18sp"  //文字大小
            android:textColor="#FF0000"   //文字颜色
            android:text="用户名："  //文字显示的内容
            android:focusable="false"/>   //是否可以获得焦点
    <!--添加一个编辑框供用户输入信息>
        <EditText
            android:id="@+id/etxt_content"
            android:layout_width="200dp"
            android:layout_height="wrap_content"
            android:gravity="center"
            android:hint="在这里输入用户名..."
            android:textSize="18sp" />
</LinearLayout>
```

 关于 Activity 类，在第 3 章我们会详细地进行介绍，这里读者只需要知道这个类会在 onCreate()方法中调用 setContentView()加载相应的布局文件即可。

- 案例分析

在 act_layout_horizontal.xml 文件中，首先定义了一个 LinearLayout 布局，其中以 android:开头的 xml 属性设置了组件的一些参数。android:layout_width 和 android:layout_height 分别设置了组件的宽度和高度，有 3 种合理的取值方式，分别如下所示。

（1）match_parent/fill_parent：设置当前 View 的大小尽可能和父控件的大小一致，在 API Level 8 以后 fill_parent 被废弃，使用 match_parent。

（2）wrap_content：设置当前 View 的大小自适应要显示的内容。

（3）固定值：设置当前 View 的尺寸为固定大小。

LinearLayout 中还设置了一个非常重要的属性——android:orientation，该属性在线性布局中也是必不可少的，其有两个取值：horizontal 和 vertical，分别指定布局中的子控件以水平和竖直的方式排列。LinearLayout 是一个视图容器，可以向其中添加组件，上述示例分别添加了一个 TextView 和一个

EditText，对于这两个控件，在 2.3.1 节会有详细介绍，这里我们只做一些简单的说明。TextView 是文本框，用于显示说明性的文字，EditText 是编辑框，可以供用户输入一些信息，例如在登录页面输入用户名和密码等。android:gravity 属性设置了 View 中内容的对齐方式，示例代码中分别指定了文字的对齐方式为左对齐和居中显示。android:textSize 设置了文字的大小，android:textColor 设置了文字的颜色，android:text 设置了文本框显示的文字。需要说明的是，Android 不提倡这种将颜色值和字符串固定写在代码中的方式，而是将其定义在资源文件中，然后通过某种方式引用，在 2.6 节我们将会专门介绍资源文件。

LinearLayout 布局相对简单，只需要把控件按顺序定义显示即可。另外，在 2.1 节中我们也提到过，在一个视图容器中不仅可以包含普通的控件，也可以包含一个视图容器。

案例 2.2　使用嵌套的线性布局

下面我们给出一个示例，运行结果如图 2.4 所示。读者可以根据运行结果去理解布局文件，用到的控件和案例 2.1 一致，注意其中 android:orientation 的值。

图 2.4　嵌套的线性布局

- **案例代码**

XML 布局文件（act_layout_linearlayout.xml）：

```
<?xml version="1.0" encoding="utf-8"?>
<LinearLayout xmlns:android="http://schemas.android.com/apk/res/android"
    android:layout_width="match_parent"
    android:layout_height="match_parent"
    android:orientation="vertical" >   //纵向线性布局
    <!-- 一个横向线性布局 -->
    <LinearLayout
        android:layout_width="match_parent"
```

```xml
        android:layout_height="wrap_content"
        android:orientation="horizontal">
    <TextView
        android:id="@+id/txt_username"
        android:layout_width="80dp"
        android:layout_height="wrap_content"
        android:gravity="right"      //文字右对齐
        android:textSize="18sp"
        android:textColor="#FF0000"
        android:text="用户名："
        android:focusable="false"/>
    <EditText
        android:id="@+id/etxt_name_content"
        android:layout_width="200dp"
        android:layout_height="wrap_content"
        android:gravity="left"     //文字左对齐
        android:hint="在这里输入用户名..."
        android:textSize="18sp" />
</LinearLayout>

<!-- 一个横向线性布局 -->
<LinearLayout
    android:layout_width="match_parent"
    android:layout_height="wrap_content"
    android:orientation="horizontal">
    <TextView
        android:id="@+id/txt_password"
        android:layout_width="80dp"
        android:layout_height="wrap_content"
        android:gravity="right"
        android:textSize="18sp"
        android:textColor="#FF0000"
        android:text="密码："
        android:focusable="false"/>
    <EditText
        android:id="@+id/etxt_pass_content"
        android:layout_width="200dp"
```

```
            android:layout_height="wrap_content"
            android:gravity="left"
            android:hint="在这里输入密码..."
            android:textSize="18sp" />
    </LinearLayout>
</LinearLayout>
```

2.2.2 相对布局

相对布局主要通过 RelativeLayout 类来实现，对于有些界面，如果很难用线性布局实现，或者使用线性布局嵌套的层次太深，可以考虑使用更灵活的相对布局。相对布局容器中子控件的位置是由父控件或者其他兄弟控件定义的。可以使当前的控件与其他控件的边界对齐，或者在某个控件的下面，又或者在父控件的中间位置。例如在图 2.1 所示的手机 QQ 的登录页面中，我们可以通过纵向线性布局去排列每一个控件，也可以通过相对布局实现，如可以设置密码输入框在账号输入框的下面，"登录"按钮在布局中水平居中。在相对布局中，子控件默认会从布局的左上角开始显示，所以，需要通过一些属性去设置控件的位置，常见的属性如表 2.1 所示。

相对布局

表 2.1 相对布局下控件常用的属性

属性	取值类型	说明
android:layout_above	其他控件 id	设置当前控件在指定 id 的控件上方
android:layout_below	其他控件 id	设置当前控件在指定 id 的控件下方
android:layout_toLeftOf	其他控件 id	设置当前控件在指定 id 的控件左侧
android:layout_toRightOf	其他控件 id	设置当前控件在指定 id 的控件右侧
android:layout_alignTop	其他控件 id	设置当前控件与指定 id 的控件上边界对齐
android:layout_alignBottom	其他控件 id	设置当前控件与指定 id 的控件下边界对齐
android:layout_alignLeft	其他控件 id	设置当前控件与指定 id 的控件左边界对齐
android:layout_alignRight	其他控件 id	设置当前控件与指定 id 的控件右边界对齐
android:layout_alignParentTop	true、false	设置当前控件是否和父布局的上方对齐
android:layout_alignParentBottom	true、false	设置当前控件是否和父布局的下方对齐
android:layout_alignParentLeft	true、false	设置当前控件是否和父布局的左边界对齐
android:layout_alignParentRight	true、false	设置当前控件是否和父布局的右边界对齐
android:layout_centerHorizontal	true、false	设置当前控件是否在父布局中水平居中
android:layout_centerVertical	true、false	设置当前控件是否在父布局中垂直居中
android:layout_centerInParent	true、false	设置当前控件是否在父布局中居中

案例 2.3 使用相对布局

下面我们通过一个示例看一下相对布局的用法，运行结果如图 2.5 所示。

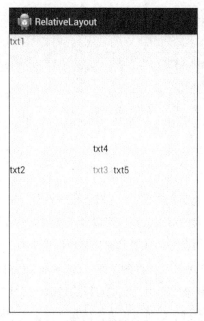

图 2.5 相对布局

- 案例代码

XML 文件（act_layout_relativelayout.xml）：

```
<?xml version="1.0" encoding="utf-8"?>
<RelativeLayout xmlns:android="http://schemas.android.com/apk/res/android"
    android:layout_width="match_parent"
    android:layout_height="match_parent" >
    <TextView
        android:id="@+id/txt1"
        android:layout_width="40dp"
        android:layout_height="40dp"
        android:textColor="#F00000"
        android:textSize="18sp"
        android:text="txt1"
        />
    <TextView
        android:id="@+id/txt2"
        android:layout_width="40dp"
        android:layout_height="40dp"
        android:layout_centerVertical="true"   //当前文本框在布局中垂直居中
        android:textColor="#0F0000"
        android:textSize="18sp"
```

```xml
            android:text="txt2"
            />
        <TextView
            android:id="@+id/txt3"
            android:layout_width="40dp"
            android:layout_height="40dp"
            android:layout_centerInParent="true"   //当前文本框在布局中处于正中间的位置
            android:textColor="#00F000"
            android:textSize="18sp"
            android:text="txt3"
            />
        <TextView
            android:id="@+id/txt4"
            android:layout_width="40dp"
            android:layout_height="40dp"
            android:layout_above="@id/txt3"   //当前文本框在txt3的上方
            android:layout_alignLeft="@id/txt3"   //当前文本框与txt3的左边界对齐
            android:textColor="#000F00"
            android:textSize="18sp"
            android:text="txt4"
            />
        <TextView
            android:id="@+id/txt5"
            android:layout_width="40dp"
            android:layout_height="40dp"
            android:layout_toRightOf="@id/txt3"   //当前文本框在txt3的右边
            android:layout_alignTop="@id/txt3"   //当前文本框与txt3的上边界对齐
            android:textColor="#0000F0"
            android:textSize="18sp"
            android:text="txt5"
            />
</RelativeLayout>
```

- **案例分析**

在图 2.5 所示界面中，第 1 个 TextView 仅仅设置了控件的大小、文字的颜色、尺寸等，控件默认显示在布局的左上角。第 2 个 TextView 设置了 android:layout_centerVertical="true"属性，控件垂直居中显示在布局中。第 3 个 TextView 设置了 android:layout_centerInParent="true"属性，控件处于布局的正中间位置。第 4 个 TextView 设置了 android:layout_above="@id/txt3"和 android:layout_ alignLeft=

"@id/txt3"属性，控件位于 txt3 的上方，且 txt4 控件的左边界与 txt3 控件的左边界对齐。类似地，第 5 个 TextView 在第 3 个 TextView 的右边，且两个控件的上边界是对齐的。

2.2.3 列表视图

列表视图-1　　列表视图-2

ListView 是 Android 中一个常用的控件，可以实现列表视图，它展示了一个垂直可滑动的下拉列表，例如图 2.6 所示的是手机中常见的文件管理页面，打开之后会有一个列表显示手机中所有的文件夹，其中的每一行称为 ListView 的一个子项。ListView 同样继承自 ViewGroup，但是和之前的 LinearLayout、RelativeLayout 不同，ListView 不是用来控制子控件的布局，而是可以根据数据源动态地添加每一个子项，需要显示的列表项由 Adapter 类提供。这样的设计也很好地符合了 MVC 原则，ListView 只负责视图的显示，而 Adapter 则提供需要显示的数据。同时，这里还用到了一个设计模式：适配器模式。Adapter 提供了将数据源转换成适合 ListView 显示的接口。ListView 有几个常用的属性，如表 2.2 所示。

图 2.6　文件列表页面

表 2.2　ListView 的相关属性

属性	属性描述
android:divider	ListView 的分隔条
android:dividerHeight	分隔条的高度
android:entries	数组资源，指定 ListView 需要显示的内容

下面我们通过几个案例说明一下 ListView 的用法。

案例 2.4　通过数组资源文件填充数据

利用列表视图可以实现如图 2.7 所示的效果。

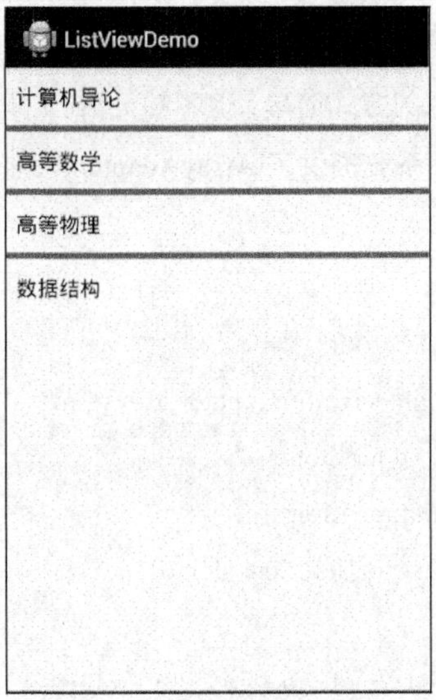

图 2.7　列表视图实例 1

- 案例代码

XML 配置文件（act_listview_demo1.xml）：

```xml
<?xml version="1.0" encoding="utf-8"?>
<LinearLayout xmlns:android="http://schemas.android.com/apk/res/android"
    android:layout_width="match_parent"
    android:layout_height="match_parent"
    android:orientation="vertical" >
    <ListView
        android:id="@+id/list"
        android:layout_width="match_parent"
        android:layout_height="wrap_content"
        android:divider="#FF0000"          //设置分隔条的颜色
        android:dividerHeight="5dp"         //设置分隔条的高度
        android:entries="@array/subjects"    //设置ListView的数据
        />
</LinearLayout>
```

- 案例分析

在布局文件中，只定义了一个 ListView 控件，ListView 设置了 android:divider 属性，定义了 ListView 中两个子项的分隔条，可以指定为固定颜色或者 Drawable 资源，这里我们重点关注 ListView 的用法，所以直接设置了一个颜色值。android:dividerHeight 设置了分隔条的高度。android:entries 设置了 ListView 需要显示的数据，这里通过 @array 引用了一个数组资源文件。资源文件在 2.6 节会有详细介绍。在 Activity 中加载上述布局即可看到案例效果。

> **案例 2.5　通过 Adapter 填充数据**

在 XML 文件中直接设置数据源的方式简单，但不够灵活。可以通过 Adapter 类为 ListView 填充数据，根据数据类型的不同，Android 系统提供了 ArrayAdapter、ListAdapter、SimpleCursorAdapter。

- 案例代码

Activity 代码：

```java
public class MainActivity extends Activity {
    protected void onCreate(Bundle savedInstanceState) {
        // TODO Auto-generated method stub
        super.onCreate(savedInstanceState);
        setContentView(R.layout.act_listview_demo1);

        ListView listView = (ListView)findViewById(R.id.list);         //获取ListView控件

        String[] data = {"计算机导论","高等数学","高等物理","数据结构"};
        //定义一个ArrayAdapter
        ArrayAdapter<String> adapter = new ArrayAdapter<String>(this, android.R.layout.simple_list_item_1, data);
        //将Adapter填充到ListView中
        listView.setAdapter(adapter);
    }
}
```

- 案例分析

布局文件是在案例 2.4 的基础上去掉 android:entries 属性，运行结果和图 2.7 所示的一致。本例中，我们首先通过 findViewById 方法得到 ListView 控件，从该方法的命名也可以看出，它是通过控件的 id 来获取控件对象的。然后定义了一个 ArrayAdapter 对象，构造方法传入了 3 个参数，第 1 个参数为一个 Context 对象，传 this 即可。第 2 个参数是一个布局文件，它定义了 ListView 每一项的外观，传入的参数是 Android 系统自带的一个布局文件，设置每个列表项都是一个普通的 TextView。第 3 个参数是一个字符串数组，定义了列表显示的数据。

> **案例 2.6　通过自定义 Adapter 填充数据，显示学生考试信息**

在实际使用 ListView 的过程中，系统已有的 Adapter 类很难满足各种各样的需求，这个时候，可以通过自定义 Adapter 填充数据。假设需要开发一个下拉列表显示学生的考试信息，每一个子项分别显示学生的姓名和考试成绩，运行结果如图 2.8 所示。

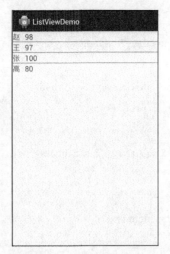

图 2.8 列表视图实例 2

- 案例代码

Activity 代码：

```
protected void onCreate(Bundle savedInstanceState) {
    super.onCreate(savedInstanceState);
    setContentView(R.layout.act_listview_demo3);

    ListView listView = (ListView)findViewById(R.id.list);    //获取ListView控件
    //一个自定义的Adapter
    CustomAdapter adapter = new CustomAdapter(this, getData());
    listView.setAdapter(adapter);    //将Adapter填充到ListView中
}

private List<Student> getData()
{
    List<Student> stuList = new ArrayList<Student>();    //构造测试数据
    stuList.add(new Student("赵", "98"));
    stuList.add(new Student("王", "97"));
    stuList.add(new Student("张", "100"));
    stuList.add(new Student("高", "80"));
    return stuList;
}
```

可以看出，Activity 的代码和案例 2.5 类似，都是给 ListView 设置了一个 Adapter，不同的是它使用了一个自定义的 Adapter：CustomAdapter。

CustomAdapter 代码：

```java
public class CustomAdapter extends BaseAdapter
{
    private List<Student> mData = new ArrayList<Student>();
    private LayoutInflater flater;

    public CustomAdapter(Context context, List<Student> list)
    {
        mData.addAll(list);
        flater = LayoutInflater.from(context);
    }
    public int getCount() {
        if(null != mData)
        {
            return mData.size();
        }
        return 0;
    }
    public Object getItem(int position) {
        if(null != mData && position < mData.size())
        {
            return mData.get(position);
        }
        return null;
    }
    @Override
    public long getItemId(int position) {
        return 0;
    }
    @Override
    public View getView(int position, View convertView, ViewGroup parent) {
        if(null == convertView)
        {
            //为每一个子项加载布局
            convertView = flater.inflate(R.layout.view_list_item, null);
        }
        //获取子item布局文件中的控件
```

```java
            TextView txtName = (TextView)convertView.findViewById(R.id.txt_name);
            TextView txtGrade = (TextView)convertView.findViewById(R.id.txt_grade);

            Student stuInfo = (Student)getItem(position);     //根据position获取数据
            if(null != stuInfo)
            {
                txtName.setText(stuInfo.getName());    //该方法可以给文本框设置显示的内容
                txtGrade.setText(stuInfo.getGrade());
            }
            return convertView;
        }
    }
```

view_list_item.xml 代码：

```xml
<?xml version="1.0" encoding="utf-8"?>
<LinearLayout xmlns:android="http://schemas.android.com/apk/res/android"
    android:layout_width="match_parent"
    android:layout_height="match_parent"
    android:orientation="horizontal" >
    <TextView
        android:id="@+id/txt_name"
        android:layout_width="30dp"
        android:layout_height="wrap_content"
        android:gravity="left"
        android:textColor="#FF0000"
        android:textSize="18sp" />
    <TextView
        android:id="@+id/txt_grade"
        android:layout_width="40dp"
        android:layout_height="wrap_content"
        android:gravity="left"
        android:textColor="#0000FF"
        android:textSize="18sp" />
</LinearLayout>
```

- **案例分析**

CustomAdapter 类继承自 BaseAdapter 类，构造函数接受两个参数，一个是 Context 对象，一个是数据源，数据源不限于 List 类型，可以是其他类型。CustomAdapter 类重写了父类的相关方法，其中最重要的是 getView 方法，系统会多次调用 getView 方法去生成 ListView 的每一项。在 getView 方法中，通过 inflate 方法加载了一个布局文件 view_list_item.xml，这个布局文件定义了 ListView 每一个子项的外观。getView 有一个入参是 position，它代表当前是 ListView 中的第几项。在代码中可以通过 position 获取对应的数据，然后设置给每一个子项中的控件。在 view_list_item.xml 布局中，定义了两个 TextView 分别用于显示学生的姓名和成绩。

2.2.4 网格视图

网格视图

GridView 也是 Android 中常用的一个组件，可以实现网格视图。GridView 和 ListView 都继承自 AbsListView，所以两者在功能和用法上都比较类似。但是网格视图是一个二维视图，例如图 2.9 所示的是手机 QQ 上的游戏中心页面，每个游戏的图标以网格的形式排列，可以上下滑动。

图 2.9 手机 QQ 游戏中心页面

和 ListView 一样，GridView 只负责视图的显示，数据源由 Adapter 提供。GridView 也有一些需要设置的属性，如表 2.3 所示。

表 2.3 GridView 的相关属性

属性	属性描述
android:columnWidth	设置列的宽度
android:numColumns	设置列数
android:verticalSpacing	每两行之间的垂直间距

续表

属性	属性描述
android:horizontalSpacing	每两列之间的水平间距
android:stretchMode	拉伸模式
android:gravity	每一格中内容的对齐方式

案例 2.7　以网格的形式排列显示 1~9 个数字

本案例我们只是简单定义一个 GridView，设置 GridView 的相关参数，说明 GridView 的用法，运行结果如图 2.10 所示。

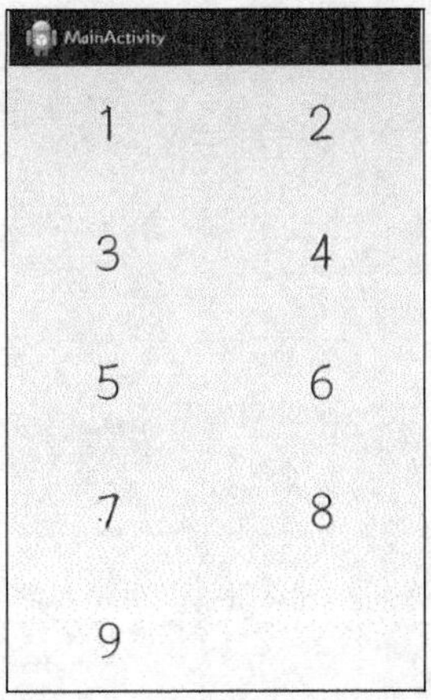

图 2.10　网格视图示例

- 案例代码

Activity 代码：

```
protected void onCreate(Bundle savedInstanceState) {
        super.onCreate(savedInstanceState);
        setContentView(R.layout.act_main);
        GridView gridView = (GridView)findViewById(R.id.gridview);    //获取GridView控件

        //创建一个自定义的Adapter
        CustomAdapter adapter = new CustomAdapter(this, getData());
        gridView.setAdapter(adapter);
}

//构造数据
private List<String> getData()
{
```

```
        List<String> data = new ArrayList<String>();
        data.add("1");
        data.add("2");
        data.add("3");
        data.add("4");
        data.add("5");
        data.add("6");
        data.add("7");
        data.add("8");
        data.add("9");
        return data;
    }
```

CustomAdapter 类代码：

```
package com.gridview.adapter;

import java.util.ArrayList;
import java.util.List;

import com.demo.gridview.R;

import android.content.Context;
import android.view.LayoutInflater;
import android.view.View;
import android.view.ViewGroup;
import android.widget.BaseAdapter;
import android.widget.ImageView;
import android.widget.TextView;

public class CustomAdapter extends BaseAdapter{
    private LayoutInflater mInflater;
    private List<String> mData = new ArrayList<String>();
    public CustomAdapter(Context context, List<String> list)
    {
        mInflater = LayoutInflater.from(context);
        mData.addAll(list);
    }
    public int getCount()
```

```java
{
    if(null == mData || mData.isEmpty())
    {
        return 0;
    }
    return mData.size();
}
public Object getItem(int position)
{
    if(null == mData || position > mData.size() - 1)
    {
        return null;
    }
    return mData.get(position);
}
public long getItemId(int position)
{
    return 0;
}
public View getView(int position, View convertView, ViewGroup parent)
{
    convertView = mInflater.inflate(R.layout.gridview_item, null);
    TextView txtvew = (TextView)convertView.findViewById(R.id.txt);
    String value = (String)mData.get(position);
    if(null != value)
    {
        txtvew.setText(value);
    }
    return convertView;
}
}
```

act_main.xml 代码:

```xml
<RelativeLayout xmlns:android="http://schemas.android.com/apk/res/android"
    xmlns:tools="http://schemas.android.com/tools"
    android:layout_width="match_parent"
    android:layout_height="match_parent"
    >
```

```
<GridView android:id="@+id/gridview"
    android:layout_width="match_parent"
    android:layout_height="wrap_content"
    android:columnWidth="120dp"          //定义列宽
    android:numColumns="auto_fit"         //定义列的个数自适应宽度和内容
    android:verticalSpacing="10dp"        //定义每两行之间的垂直间距
    android:horizontalSpacing="10dp"      //定义每两列之间的垂直间距
    android:gravity="center"/>
</RelativeLayout>
```

- 案例分析

从代码中可以看到，GridView 的使用方法和 ListView 非常类似，都是通过 setAdapter 方法给视图设置数据源，Adapter 的自定义方式与 2.2.3 节介绍的也一致。

2.3 常用控件

我们在 2.1 节中介绍过控件的概念，它们是组成 Android 界面的基本元素。无论多么复杂美观的应用界面，都可以通过组合基本控件实现。本节我们将介绍一些常用的控件。

2.3.1 文本框和编辑框

文本框通过 TextView 控件实现，用于文字的显示。编辑框通过 EditText 实现，它继承自 TextView，属性和用法与 TextView 一致，只不过它允许用户改变其中的内容。文本框的属性如表 2.4 所示。

基本控件之文本框和编辑框、按钮、图片控件

表 2.4 文本框的属性

属性	属性描述
android:text	文本框显示的文字
android:textSize	显示文字的大小
android:textColor	显示文字的颜色
android:gravity	文字在文本框中的位置
android:ellipsize	文字内容超过文本框大小时的显示方式
android:password	是否以点代替显示输入的文字
android:editable	文本框是否可编辑
android:hint	当文本框的内容为空时，显示的提示文字
android:singleLine	是否单行显示
android:autoLink	是否将指定格式的文本转化为可单击的链接
android:cursorVisible	光标是否可见
android:drawableLeft	在文本框中文本的左侧显示指定图片

表 2.4 中有两个属性的取值需要特别说明一下。

（1）android:ellipsize 的取值如下所示。

none：文字超长不做任何处理。

start：在文字的起始处显示省略号。

middle：在文字的中间显示省略号。

end：在文字的结尾处显示省略号。

marquee：文字滚动显示。

（2）android:autoLink 的取值如下所示。

none：不进行文本检测。

web：将文本框中的网址转换为链接。

email：将文本框中的邮箱地址转换为链接。

phone：将文本框中的电话号码转换为链接。

map：将文本框中的地址转换为链接。

all：等价于设置为 web|email|phone|map。

下面我们通过几个实例来介绍一下文本框和编辑框的使用。

案例 2.8　显示不同颜色、大小和不同位置的文字

本例配置文件中定义了 3 个文本框，设置了 3 种不同的文字颜色和文字对齐方式，字体大小逐渐增大，运行结果如图 2.11 所示。

图 2.11　不同颜色和大小的文本框

- 案例代码

XML 配置文件（act_txt1.xml）：

```
<LinearLayout xmlns:android="http://schemas.android.com/apk/res/android"
    xmlns:tools="http://schemas.android.com/tools"
```

```xml
    android:layout_width="match_parent"
    android:layout_height="match_parent"
    android:orientation="vertical"
    >
    <TextView
        android:id="@+id/txt1"
        android:layout_width="match_parent"
        android:layout_height="wrap_content"
        android:gravity="left"      //文字靠左显示
        android:textSize="18sp"     //文字大小
        android:textColor="#FF0000"     //文字颜色为红色
        android:text="Hello World"
        />
    <TextView
        android:id="@+id/txt2"
        android:layout_width="match_parent"
        android:layout_height="wrap_content"
        android:gravity="center"    //文字居中显示
        android:textSize="28sp"     //文字大小
        android:textColor="#00FF00"     //文字颜色为绿色
        android:text="Hello World"
        />
    <TextView
        android:id="@+id/txt3"
        android:layout_width="match_parent"
        android:layout_height="wrap_content"
        android:gravity="right"     //文字靠右显示
        android:textSize="38sp"     //文字大小
        android:textColor="#0000FF"     //文字颜色为蓝色
        android:text="Hello World"
        />
</LinearLayout>
```

案例 2.9　文字超长时的处理

本例配置文件中定义了 4 种按不同的属性值显示的超长文本，第 1 个文本框指定在文字起始处显示省略号，第 2 个文本框指定在文字的中间显示省略号，第 3 个文本框指定在文字的结尾处显示省略号，第 4 个文本框指定文字滚动显示。另外设置了属性 android:singleLine=true，控制文本单行显示，否则超长的文本会自动换行显示。运行结果如图 2.12 所示。

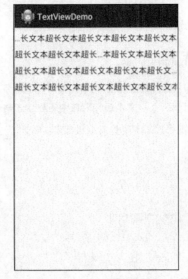

图 2.12　超长文本的处理

- **案例代码**

XML 配置文件（act_txt2.xml）：

```xml
<LinearLayout xmlns:android="http://schemas.android.com/apk/res/android"
    xmlns:tools="http://schemas.android.com/tools"
    android:layout_width="match_parent"
    android:layout_height="match_parent"
    android:orientation="vertical"
    >
    <TextView
        android:id="@+id/txt1"
        android:layout_width="match_parent"
        android:layout_height="wrap_content"
        android:layout_marginTop="10dp"
        android:textSize="18sp"
        android:textColor="#0000FF"
        android:singleLine="true"    //单行显示
        android:ellipsize="start"    //文本超长时，在起始处显示省略号
        android:text="超长文本超长文本超长文本超长文本超长文本超长文本"
        />
    <TextView
        android:id="@+id/txt2"
```

```
            android:layout_width="match_parent"
            android:layout_height="wrap_content"
            android:layout_marginTop="10dp"
            android:textSize="18sp"
            android:textColor="#0000FF"
            android:singleLine="true"      //单行显示
            android:ellipsize="middle"     //文本超长时在中间显示省略号
            android:text="超长文本超长文本超长文本超长文本超长文本超长文本"
            />
    <TextView
            android:id="@+id/txt3"
            android:layout_width="match_parent"
            android:layout_height="wrap_content"
            android:layout_marginTop="10dp"
            android:textSize="18sp"
            android:textColor="#0000FF"
            android:singleLine="true"      //单行显示
            android:ellipsize="end"        //文本超长时在结尾处显示省略号
            android:text="超长文本超长文本超长文本超长文本超长文本超长文本"
            />
    <TextView
            android:id="@+id/txt4"
            android:layout_width="match_parent"
            android:layout_height="wrap_content"
            android:layout_marginTop="10dp"
            android:textSize="18sp"
            android:textColor="#0000FF"
            android:focusable="true"
            android:singleLine="true"      //单行显示
            android:ellipsize="marquee"    //文本超长时滚动显示
            android:marqueeRepeatLimit="marquee_forever"   //设置滚动次数为一直滚动
            android:text="超长文本超长文本超长文本超长文本超长文本超长文本"
            />
</LinearLayout>
```

案例 2.10　将指定格式的文本转化为可单击的链接

本例配置文件分别定义了 3 个文本框，分别将其中的网址、邮箱地址、电话号码转化为可以单击的超链接，运行结果如图 2.13 所示。

图 2.13 将指定格式的文本转换为可单击的超链接

- 案例代码

XML 配置文件（act_txt3.xml）：

```xml
<LinearLayout xmlns:android="http://schemas.android.com/apk/res/android"
    xmlns:tools="http://schemas.android.com/tools"
    android:layout_width="match_parent"
    android:layout_height="match_parent"
    android:orientation="vertical"
    >
<TextView
        android:id="@+id/txt1"
        android:layout_width="match_parent"
        android:layout_height="wrap_content"
        android:layout_marginTop="10dp"
        android:textSize="18sp"
        android:textColor="#0000FF"
        android:autoLink="web"
        android:text="人邮的网址是：http://www.ptpress.com.cn"
        />
<TextView
        android:id="@+id/txt2"
        android:layout_width="match_parent"
        android:layout_height="wrap_content"
        android:layout_marginTop="10dp"
```

            android:textSize="18sp"

            android:textColor="#0000FF"

            android:autoLink="email"

            android:text="我的邮箱是：123@126.com"

         />

    <TextView

            android:id="@+id/txt1"

            android:layout_width="match_parent"

            android:layout_height="wrap_content"

            android:layout_marginTop="10dp"

            android:textSize="18sp"

            android:textColor="#0000FF"

            android:autoLink="phone"

            android:text="我的联系方式是：12345678910"

         />

</LinearLayout>

2.3.2 按钮

按钮通过 Button 控件实现，Button 类继承自 TextView，它可以供用户单击，当用户单击之后，就会触发一个 onClick 事件，可以通过监听 onClick 事件做一些自定义的处理。图 2.1 所示的手机 QQ 登录界面的"登录"按钮就是一个 Button 控件。

案例 2.11　切换"Hello"和"World"的显示

下面我们通过一个实例介绍一下 Button 的使用，运行结果如图 2.14 所示，左图是初始显示，右图是单击"切换"按钮后的显示。

图 2.14　Button 控件示例

- **案例代码**

XML 配置文件（act_btn.xml）：

```xml
<?xml version="1.0" encoding="utf-8"?>
<LinearLayout xmlns:android="http://schemas.android.com/apk/res/android"
    android:layout_width="match_parent"
    android:layout_height="match_parent"
    android:orientation="horizontal" >
    <TextView
        android:id="@+id/txt"
        android:layout_width="100dp"
        android:layout_height="50dp"
        android:textSize="18sp"
        android:textColor="#0000FF"
        android:gravity="center"
        android:text="Hello"/>
    <Button
        android:id="@+id/btn"
        android:layout_width="100dp"
        android:layout_height="50dp"
        android:textSize="18sp"
        android:textColor="#0000FF"
        android:gravity="center"
        android:text="切换"/>
</LinearLayout>
```

Activity 代码（MainActivity.java）：

```java
public class MainActivity extends Activity {
    protected void onCreate(Bundle savedInstanceState) {
        super.onCreate(savedInstanceState);
        setContentView(R.layout.act_btn);
        initWidget();
    }
    private void initWidget()
    {
        Button btn = (Button)findViewById(R.id.btn);         //获取Button控件
        btn.setOnClickListener(new OnClickListener() {       //为Button设置单击事件监听器
            public void onClick(View v) {
                TextView txt = (TextView)findViewById(R.id.txt);        //获取TextView控件
```

```
                    String str = txt.getText().toString();    //获取TextView当前显示的文字
                    //切换显示"Hello"和"World"
                    txt.setText("Hello".equals(str) ? "World" : "Hello");
                }
            });
        }
    }
```

- 案例分析

本例配置文件中定义了一个 TextView 和 Button，可以看到，Button 控件的属性设置和 TextView 一致。

在 Activity 代码中，首先通过 findViewById()方法获取了 Button 控件，然后为 Button 控件设置了一个监听事件，监听用户的单击操作。当用户进行单击后，对界面文本框中的文字进行判断，切换"Hello"和"World"的显示。

2.3.3 单选按钮和复选框

在有些界面中，信息并不一定完全需要用户输入，可以提供一组信息让用户进行选择，这可以通过单选按钮和复选框实现。单选按钮和复选框分别通过 RadioButton 和 CheckBox 实现，两者都继承自 Button，因此它们可以使用 Button 控件的属性和方法。相比于普通的按钮，它们多了一个"选中"的概念。

案例 2.12　选择性别与爱好

下面我们通过一个实例介绍一下单选按钮和复选框的用法，运行结果如图 2.15 所示。

图 2.15　单选按钮和复选框

- 案例代码

配置文件（act_main.xml）：

```xml
<?xml version="1.0" encoding="utf-8"?>
<LinearLayout xmlns:android="http://schemas.android.com/apk/res/android"
    android:layout_width="match_parent"
    android:layout_height="match_parent"
    android:orientation="vertical" >
    <RadioGroup
        android:id="@+id/radioGroup"
        android:layout_width="300dp"
        android:layout_height="50dp"
        android:gravity="center_vertical"
        android:orientation="horizontal" >
        <RadioButton
            android:id="@+id/radio_male"
            android:layout_width="wrap_content"
            android:layout_height="wrap_content"
            android:checked="true"
            android:text="男"
            android:textSize="18sp" />
        <RadioButton
            android:id="@+id/radio_female"
            android:layout_width="wrap_content"
            android:layout_height="wrap_content"
            android:text="女"
            android:textSize="18sp" />
    </RadioGroup>
    <LinearLayout
        android:layout_width="match_parent"
        android:layout_height="wrap_content"
        android:orientation="horizontal" >
        <CheckBox
            android:id="@+id/checkbox_run"
            android:layout_width="wrap_content"
            android:layout_height="wrap_content"
            android:text="跑步"
```

```xml
            android:textSize="18sp" />
        <CheckBox
            android:id="@+id/checkbox_swim"
            android:layout_width="wrap_content"
            android:layout_height="wrap_content"
            android:text="游泳"
            android:textSize="18sp" />
        <CheckBox
            android:id="@+id/checkbox_reading"
            android:layout_width="wrap_content"
            android:layout_height="wrap_content"
            android:text="读书"
            android:textSize="18sp" />
    </LinearLayout>
</LinearLayout>
```

Activity 代码：

```java
public class MainActivity extends Activity {

    private String mRadioValue = "";
    private List<String> mCheckBoxValue = new ArrayList<String>();

    protected void onCreate(Bundle savedInstanceState) {
        super.onCreate(savedInstanceState);
        setContentView(R.layout.act_main);
        initWidget();
    }

    private void initWidget()
    {
        //获取RadioGroup控件
        RadioGroup radioGroup = (RadioGroup) findViewById(R.id.radioGroup);
        radioGroup.setOnCheckedChangeListener(new RadioGroup.OnCheckedChangeListener() { // 为RadioGroup控件设置监听事件
            public void onCheckedChanged(RadioGroup group, int checkedId) {
                int radioId = group.getCheckedRadioButtonId();
                switch (radioId) {
```

```java
                case R.id.radio_male:        //选中的按钮是"男"
                    mRadioValue = "男";
                    break;
                case R.id.radio_female:      //选中的按钮是"女"
                    mRadioValue = "女";
                    break;
                default:
                    mRadioValue = "男";
                    break;
            }
        }
    });
    CheckBox checkRun = (CheckBox) findViewById(R.id.checkbox_run);      //获取复选框控件
    CheckBox checkReading = (CheckBox) findViewById(R.id.checkbox_reading);
    CheckBox checkSwim = (CheckBox) findViewById(R.id.checkbox_swim);
    checkRun.setOnCheckedChangeListener(checkedListener);     //为复选框控件设置监听事件
    checkReading.setOnCheckedChangeListener(checkedListener);
    checkSwim.setOnCheckedChangeListener(checkedListener);
}
OnCheckedChangeListener checkedListener = new OnCheckedChangeListener() {
    public void onCheckedChanged(CompoundButton buttonView,
            boolean isChecked) {
        switch (buttonView.getId()) {
        case R.id.checkbox_run: {        //单击的是"跑步"
            if (isChecked) {    //检测选中状态
                mCheckBoxValue.add("run");    //如果选中了，将值添加到集合中
            } else {
                mCheckBoxValue.remove("run");   //如果取消选中，将值移除
            }
            break;
        }
        case R.id.checkbox_reading: {    //单击的是"读书"
            if (isChecked) {
                mCheckBoxValue.add("reading");
            } else {
                mCheckBoxValue.remove("reading");
```

```
                }
                break;
            }
            case R.id.checkbox_swim: {     //单击的是"游泳"
                if (isChecked) {
                    mCheckBoxValue.add("swimming");
                } else {
                    mCheckBoxValue.remove("swimming");
                }
                break;
            }
            default:
                break;
            }
        }
    };
}
```

- **案例分析**

本例配置文件首先定义了一个 RadioGroup 控件,在其中放置 RadioButton 控件,这样可以保证"男"和"女"两个按钮只被选择一次。接着定义了一个 LinearLayout 控件,方向为水平布局,在其中放置了 3 个复选框。

在 Activity 中,分别获取了单选框和复选框控件,然后为其设置监听事件,并获取相应的信息。

2.3.4 图片控件

图片控件通过 ImageView 实现,它主要用于显示图片。ImageView 的使用方法也比较简单,表 2.5 列出了 ImageView 一些常见的属性。

表 2.5 ImageView 常见的属性

属性	属性描述
android:adjustViewBounds	设置 ImageView 控件是否调整自己的边界保持所显示图片的长宽比例
android:maxHeight	ImageView 控件的最大高度
android:maxWidth	ImageView 控件的最大宽度
android:scaleType	设置图片如何调整自己的大小去适应 ImageView 控件的大小
android:src	设置 ImageView 显示的 Drawable 对象

其中,android:scaleType 的取值如表 2.6 所示。

表 2.6　android:scaleType 的取值

取值	说明
matrix	默认的显示方式，不改变图片的大小，从 ImageView 的左上角开始显示，超出部分裁剪掉
fitXY	对图片横向、纵向缩放，使得图片填满整个 ImageView 显示
fitStart	保持图片的纵横比进行缩放，直至图片较长的一边和 ImageView 对应的边相等，然后显示在 ImageView 的左上部分
fitCenter	保持图片的纵横比进行缩放，直至图片较长的一边和 ImageView 对应的边相等，图片居中显示
fitEnd	保持图片的纵横比进行缩放，直至图片较长的一边和 ImageView 对应的边相等，然后显示在图片的右下部分
center	保持原图的大小，将图片显示在 ImageView 的中间，超出部分裁剪掉
centerCrop	原图小于 ImageView 时，保持图片的纵横比放大，直至图片填满整个 ImageView，超出部分裁剪掉
centerInside	保持图片的纵横比进行缩放，直至原图完全显示在 ImageView 中

下面我们通过示例了解一下 ImageView 的用法，并直观地感受一下 android:scaleType 属性不同取值对图片的处理。

案例 2.13　图片尺寸大于 ImageView 控件尺寸的大小

定义 ImageView 控件的大小为 200dp×200dp，分别选取两张图片进行展示。图 2.16 所示的是待展示的图片，左边图片大于 ImageView 控件的大小，右边图片小于 ImageView 控件的大小。图 2.17 所示为 ImageView 的展示效果，为了对比图片在 ImageView 中的位置，我们将 ImageView 的背景颜色设置为蓝色。

图 2.16　示例图片

图 2.17　ImageView 的运行结果

- **案例代码**

XML 配置文件（act_main.xml）：

```
<LinearLayout xmlns:android="http://schemas.android.com/apk/res/android"
    xmlns:tools="http://schemas.android.com/tools"
    android:layout_width="match_parent"
    android:layout_height="match_parent"
    android:orientation="vertical" >

    <LinearLayout
        android:layout_width="match_parent"
        android:layout_height="wrap_content"
        android:gravity="center"
        android:orientation="horizontal">
        <ImageView
            android:id="@+id/img1"
            android:layout_width="120dp"
            android:layout_height="120dp"
            android:scaleType="matrix"
            android:background="#0000FF"
            android:src="@drawable/pic1"
            />
        <ImageView
            android:id="@+id/img2"
```

```xml
            android:layout_width="120dp"
            android:layout_height="120dp"
            android:layout_marginLeft="20dp"
            android:scaleType="fitXY"
            android:background="#0000FF"
            android:src="@drawable/pic1"
            />
    </LinearLayout>

    <LinearLayout
        android:layout_width="match_parent"
        android:layout_height="wrap_content"
        android:layout_marginTop="20dp"
        android:gravity="center"
        android:orientation="horizontal">
        <ImageView
            android:id="@+id/img3"
            android:layout_width="120dp"
            android:layout_height="120dp"
            android:scaleType="fitStart"
            android:background="#0000FF"
            android:src="@drawable/pic1"
            />
        <ImageView
            android:id="@+id/img4"
            android:layout_width="120dp"
            android:layout_height="120dp"
            android:layout_marginLeft="20dp"
            android:scaleType="fitCenter"
            android:background="#0000FF"
            android:src="@drawable/pic1"
            />
    </LinearLayout>

    <LinearLayout
        android:layout_width="match_parent"
        android:layout_height="wrap_content"
```

```xml
        android:layout_marginTop="20dp"
        android:gravity="center"
        android:orientation="horizontal">
    <ImageView
        android:id="@+id/img5"
        android:layout_width="120dp"
        android:layout_height="120dp"
        android:scaleType="fitEnd"
        android:background="#0000FF"
        android:src="@drawable/pic1"
        />
    <ImageView
        android:id="@+id/img6"
        android:layout_width="120dp"
        android:layout_height="120dp"
        android:layout_marginLeft="20dp"
        android:scaleType="center"
        android:background="#0000FF"
        android:src="@drawable/pic1"
        />
</LinearLayout>

<LinearLayout
    android:layout_width="match_parent"
    android:layout_height="wrap_content"
    android:layout_marginTop="20dp"
    android:gravity="center"
    android:orientation="horizontal">
    <ImageView
        android:id="@+id/img7"
        android:layout_width="120dp"
        android:layout_height="120dp"
        android:scaleType="centerCrop"
        android:background="#0000FF"
        android:src="@drawable/pic1"
        />
    <ImageView
```

```
                android:id="@+id/img8"
                android:layout_width="120dp"
                android:layout_height="120dp"
                android:layout_marginLeft="20dp"
                android:scaleType="centerInside"
                android:background="#0000FF"
                android:src="@drawable/pic1"
                />
    </LinearLayout>
</LinearLayout>
```

2.3.5 进度条和拖动条

进度条和拖动条

在 Android 应用中，我们经常会用到进度条和拖动条。进度条可以用来显示当前操作的进度，拖动条在进度条的基础上做了扩展，允许用户随意拖动当前的进度，例如在音乐或视频播放器中，用户可以拖动滑块实现快进或快退。进度条通过 ProgressBar 实现，拖动条通过 SeekBar 实现。表 2.7 说明了 ProgressBar 一些常见的属性。

表 2.7 ProgressBar 的属性

属性	属性说明
android:max	进度条的最大值
android:maxHeight	进度条的最大高度
android:maxWidth	进度条的最大宽度
android:minHeight	进度条的最小高度
android:minWidth	进度条的最小宽度
android:progress	进度条默认显示的进度

下面我们通过一个实例说明一下进度条的使用方法。

- 案例 2.14 使用进度条

本例在 XML 配置文件中定义了一个 progressBar 对象，设置进度条的样式为水平显示，进度条的最大值设置为 100，然后在 Activity 中获取 ProgressBar 控件，每隔 1s 调用 setProgress()函数刷新一次进度显示，运行结果如图 2.18 所示。

- 案例代码

Activity 代码：
```
public class MainActivity extends Activity {
    private int progress = 0;
```

```java
private ProgressBar progressBar;     //定义一个进度条对象
private Handler mHandler = new Handler();
private Runnable task = new Runnable()    //定义一个线程,不断更新当前进度
{
    public void run() {
        ++progress;     //进度自增
        progressBar.setProgress(progress);     //设置进度条当前显示的进度
        if(progress >= 100)    //进度条已达到最大值
        {
            mHandler.removeCallbacks(this);
        }else
        {
            mHandler.postDelayed(this, 1000);
        }
    }
};
protected void onCreate(Bundle savedInstanceState) {
    super.onCreate(savedInstanceState);
    setContentView(R.layout.activity_main);
    progressBar = (ProgressBar)findViewById(R.id.progress);    //获取进度条控件
    mHandler.postDelayed(task, 1000);     //每隔1s刷新一次进度条
}
```

图 2.18 进度条

xml 配置文件：

```
<RelativeLayout xmlns:android="http://schemas.android.com/apk/res/android"
    xmlns:tools="http://schemas.android.com/tools"
    android:layout_width="match_parent"
    android:layout_height="match_parent"
    >
    <ProgressBar     //定义一个进度条控件
        android:id="@+id/progress"
        android:layout_width="match_parent"
        android:layout_height="5dp"
        android:layout_marginTop="10dp"
        android:layout_marginLeft="10dp"
        android:layout_marginRight="10dp"
        android:max="100"      //进度条最大值设置为100
        android:progress="0"     //进度条当前默认显示设置为0
        style="@android:style/Widget.ProgressBar.Horizontal"/>     //进度条样式设置为水平
</RelativeLayout>
```

案例 2.15　使用拖动条

SeekBar 和进度条类似，只不过多了一个滑动块可供用户拖动，SeekBar 可以通过属性 android:thumb 来改变滑块的显示。下面我们通过一个实例说明一下 SeekBar 的使用方法，运行结果如图 2.19 所示。

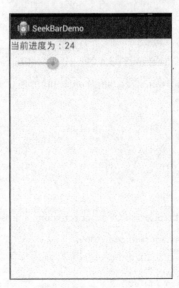

图 2.19　拖动条

- **案例代码**

Activity 代码：

```java
public class MainActivity extends Activity {
    protected void onCreate(Bundle savedInstanceState) {
        super.onCreate(savedInstanceState);
        setContentView(R.layout.activity_main);
        initWidget();
    }
    private void initWidget()
    {
        final TextView txtShow = (TextView)findViewById(R.id.txt_show);
        SeekBar seekBar = (SeekBar)findViewById(R.id.seekbar);
        //为拖动条添加监听事件
        seekBar.setOnSeekBarChangeListener(new OnSeekBarChangeListener()
        {
            //拖动产生进度改变时回调该方法
            public void onProgressChanged(SeekBar seekBar, int progress,
                    boolean fromUser) {
                txtShow.setText("当前进度为: " + progress);      //进度变化时显示当前进度
            }
            //刚开始拖动时回调该方法
            public void onStartTrackingTouch(SeekBar seekBar) {
                // TODO Auto-generated method stub
            }
            //拖动停止时回调该方法
            public void onStopTrackingTouch(SeekBar seekBar) {
                // TODO Auto-generated method stub
            }
        });
    }
}
```

xml 配置文件：

```xml
<LinearLayout xmlns:android="http://schemas.android.com/apk/res/android"
    xmlns:tools="http://schemas.android.com/tools"
    android:layout_width="match_parent"
    android:layout_height="match_parent"
    android:orientation="vertical" >
    <TextView
        android:id="@+id/txt_show"
```

```
            android:layout_width="match_parent"
            android:layout_height="wrap_content"
            android:textSize="20sp"/>
    <SeekBar //定义一个进度条
            android:id="@+id/seekbar"
            android:layout_width="match_parent"
            android:layout_height="wrap_content"
            android:layout_marginTop="10dp"
            android:progress="10"
            android:max="100"/>
</LinearLayout>
```

- 案例分析

从配置文件中可以看到，拖动条的配置与进度条类似。在 Activity 中，需要调用 setOnSeekBarChangeListener()为拖动条设置监听事件，当有拖动事件产生时，系统会回调当前拖动的进度，可以在 onProgressChanged()方法中获取当前进度并做相应的处理。

2.4 对话框

对话框也是一种比较常见的用于交互的控件，是提示用户做出决定或者输入额外信息的窗口，一般不会填充整个屏幕，需要用户采取进一步操作之后才能继续执行。Android 中对话框的基类是 Dialog，一般使用其子类 AlertDialog、ProgressDialog 等。其中，AlertDialog 功能最丰富，应用最广泛。

对话框

2.4.1 简单对话框

AlertDialog 提供了一些方法用于生成带消息和操作按钮的对话框，对话框的内容还可以是列表或者是自定义的 View。对话框可通过 AlertDialog.Builder 进行一些设置，AlertDialog.Builder 类支持的方法如表 2.8 所示。

表 2.8 AlertDialog.Builder 支持的方法

方法名	方法说明
create()	创建一个 AlertDialog 对话框
setCancelable(boolean cancelable)	设置当前对话框是否可以被取消
setIcon(Drawable icon)	设置对话框的标题图标
setItems(CharSequence[] items, DialogInterface.OnClickListener listener)	将对话框的内容设置为列表
setMessage(CharSequence message)	设置对话框显示的消息

续表

方法名	方法说明
setNegativeButton(CharSequence text, DialogInterface.OnClickListener listener)	设置"取消"按钮的显示和事件处理
setPositiveButton(CharSequence text, DialogInterface.OnClickListener listener)	设置"确定"按钮的显示和事件处理
setTitle(CharSequence title)	设置对话框显示的标题
show()	显示对话框
setView(View view)	将对话框的内容区域设置为自定义的 View

下面我们通过一个实例说明一下 AlertDialog 的使用。

案例 2.16　使用简单对话框

本例是在 Activity 中显示一个按钮，单击按钮之后，使用 AlertDialog.Builder 创建一个对话框并显示。对话框设置了标题、消息和两个按钮，运行结果如图 2.20 所示。

图 2.20　简单对话框

- **案例代码**

Activity 代码：

```
public class MainActivity extends Activity {
    protected void onCreate(Bundle savedInstanceState) {
        super.onCreate(savedInstanceState);
```

```java
setContentView(R.layout.activity_main);
Button btn = (Button)findViewById(R.id.btn_dialog);    //获取布局中的Button按钮
//定义一个AlertDialog.Builder对象
final AlertDialog.Builder builder = new AlertDialog.Builder(this);
btn.setOnClickListener(new OnClickListener() {
    public void onClick(View v) {
        builder.setTitle("Dialog");      //设置对话框的标题
        builder.setMessage("This is a Dialog");  //设置对话框显示的消息内容
        //为对话框添加按钮
        builder.setPositiveButton("OK", new DialogInterface.OnClickListener()
        {
            public void onClick(DialogInterface dialog, int which) {
                Toast.makeText(MainActivity.this, "press OK", Toast.LENGTH_ LONG).show();
            }
        });
        //为对话框添加按钮
        builder.setNegativeButton("Cancel", new DialogInterface.OnClick Listener()
        {
            public void onClick(DialogInterface dialog, int which) {
                Toast.makeText(MainActivity.this, "press Cancel", Toast.LENGTH_LONG).show();
            }
        });
        builder.create().show();    //创建对话框并显示
    }
});
}
}
```

2.4.2 列表对话框

AlertDialog 除了可以创建简单的对话框之外，还可以创建列表对话框，调用 AlertDialog.Builder 对应的 setXXXItems()方法可以创建简单的列表、带有单选按钮的列表、带有多选框的列表。

案例 2.17　使用列表对话框选择语言

下面我们以带有多选框的列表内容为例，创建一个列表对话框，运行结果如图 2.21 所示。

图 2.21 列表对话框

- **案例代码**

Activity 代码:

```java
public class MainActivity extends Activity {
    private String[] lan = {"C++", "JAVA", "Python", "Ruby", "PHP"};     //定义列表显示的内容
    protected void onCreate(Bundle savedInstanceState) {
        super.onCreate(savedInstanceState);
        setContentView(R.layout.activity_main);
        Button btn = (Button)findViewById(R.id.btn_dialog);
        //创建一个AlertDialog.Builder对象
        final AlertDialog.Builder builder = new AlertDialog.Builder(this);
        btn.setOnClickListener(new OnClickListener() {
            public void onClick(View v) {
                builder.setTitle("langage");
                //设置列表的内容
                builder.setMultiChoiceItems(lan, null, new OnMultiChoiceClickListener(){
                    public void onClick(DialogInterface dialog, int which,
                        boolean isChecked) {
                        if(isChecked)      //判断是否被选中
                        {
                            Toast.makeText(MainActivity.this, "you have choice:" + lan[which], Toast.LENGTH_LONG).show();
                        }
                    }
                });
            }
        });
```

```
                    builder.create().show();        //创建对话框并显示
                }
            });
        }
    }
```

- **案例分析**

本例在 Activity 中显示了一个按钮,单击按钮之后,创建了一个支持多选的列表对话框。创建的方式是调用 AlertDialog.Builder 的 setMultiChoiceItems(CharSequence[] items, boolean[] checkedItems, DialogInterface.OnMultiChoiceClickListener listener),第 1 个参数是列表要显示的内容,第 2 个参数是设置列表的默认选择情况,第 3 个参数是设置一个监听事件。当列表的某一项被单击后,系统就会回调 onClick(DialogInterface dialog,int which,boolean isChecked)方法,其中,which 表示用户单击了第几项,isChecked 表示用户单击后,当前项是否处于选中状态。

2.4.3 自定义对话框

除了创建已有的对话框样式外,AlertDialog.Builder 还支持调用 serView()方法显示自定义的 View,使用方法比较简单,下面我们通过一个实例具体说明一下。

案例 2.18 使用自定义对话框制作登录页面

本例在 Activity 中显示一个按钮,单击按钮之后,创建一个自定义布局的对话框,包括一个用户名输入框和一个密码输入框,用于做登录操作。运行结果如图 2.22 所示。

图 2.22 自定义对话框

- 案例代码

Activity 代码：

```java
public class MainActivity extends Activity {
    protected void onCreate(Bundle savedInstanceState) {
        super.onCreate(savedInstanceState);
        setContentView(R.layout.activity_main);
        Button btn = (Button)findViewById(R.id.btn_dialog);
        //创建一个AlertDialog.Builder对象
        final AlertDialog.Builder builder = new AlertDialog.Builder(this);
        btn.setOnClickListener(new OnClickListener() {
            public void onClick(View v) {
                builder.setTitle("Login");
                LayoutInflater inflater = LayoutInflater.from(MainActivity.this);
                View view = inflater.inflate(R.layout.vew_login, null);   //加载一个布局
                builder.setView(view);      //将自定义的布局设置在对话框中
                final EditText editUserName = (EditText)view.findViewById (R.id.edit_username);
                final EditText editPassword = (EditText)view.findViewById(R.id.edit_ password);
                builder.setPositiveButton("Login", new DialogInterface.OnClick Listener()
                {
                    public void onClick(DialogInterface dialog, int which) {
                        String username = editUserName.getText(). toString();
                        String password = editPassword.getText(). toString();
                    }
                });
                builder.setNegativeButton("Cancel", new DialogInterface.OnClick Listener()
                {
                    public void onClick(DialogInterface dialog, int which) {
                        Toast.makeText(MainActivity.this, "press Cancel", Toast.LENGTH_LONG).show();
                    }
                });
                builder.create().show();
            }
        });
    }
}
```

view_login.xml 文件：

```xml
<?xml version="1.0" encoding="utf-8"?>
<LinearLayout xmlns:android="http://schemas.android.com/apk/res/android"
    android:layout_width="match_parent"
    android:layout_height="match_parent"
    android:orientation="vertical"
    android:gravity="center_horizontal"
    android:padding="10dp">
    <EditText
        android:id="@+id/edit_username"
        android:layout_width="match_parent"
        android:layout_height="wrap_content"
        android:textSize="18sp"
        android:hint="username"/>
    <EditText
        android:id="@+id/edit_password"
        android:layout_width="match_parent"
        android:layout_height="wrap_content"
        android:textSize="18sp"
        android:hint="password"/>
</LinearLayout>
```

2.5 菜单

菜单在应用中也是很常见的一个组件,它能为用户提供更多的操作。Android 中的菜单分为选项菜单和上下文菜单,选项菜单一般是应用的主菜单项,在应用的任何地方按菜单键都会弹出来。上下文菜单是指在应用中的某个地方长按会弹出的菜单,类似于在电脑上单击鼠标右键弹出来的菜单。如图 2.23 所示,左边是微信的选项菜单,在微信的任何页面按菜单键都会弹出。右边是微信的上下文菜单,在微信的某条聊天记录上长按会弹出该菜单。

菜单

2.5.1 选项菜单

无论是选项菜单还是上下文菜单,菜单项都可以在 XML 文件中定义,这样可以将菜单的视图和操作分开,使代码的结构看起来更清晰。菜单的 XML 文件放在项目的 res/menu 目录下,使用的元素为<menu>和<item>。示例代码如下:

```xml
<?xml version="1.0" encoding="utf-8"?>
<menu xmlns:android="http://schemas.android.com/apk/res/android">
```

```
        <item android:id="@+id/menu_more"
            android:title="更多"
            />
        <item android:id="@+id/menu_help"
            android:title="帮助" />
</menu>
```

图 2.23　微信的选项菜单和上下文菜单

<menu/>元素定义了一个菜单，其中可以包括一个或多个<item>元素，每个<item>元素定义了菜单中的某一项，<item>的 android:id 属性标识了当前菜单项，android:title 属性设置了菜单项显示的文字。另外，<item>还可以嵌套<menu>元素作为子菜单，例如：

```
<?xml version="1.0" encoding="utf-8"?>
<menu xmlns:android="http://schemas.android.com/apk/res/android">
    <item android:id="@+id/oper"
        android:title="操作" >
        <!-- 子菜单 -->
        <menu>
            <item android:id="@+id/menu_create"
                android:title="新建" />
            <item android:id="@+id/menu_open"
                android:title="打开" />
        </menu>
    </item>
</menu>
```

要在 Activity 中创建选项菜单，需要重写 Activity 的 onCreateOptionsMenu()方法，在该方法中，将定义的菜单加入到 Menu 对象中。当单击选项菜单中的某一项时，Activity 会回调 onOptionsItemSelected(MenuItem item)，其中 MenuItem 就代表选项菜单中的某一项。下面我们通过一个具体实例看一下选项菜单的创建与使用。

案例 2.19　制作"添加""删除""查询"和"退出"选项菜单

在 XML 文件中定义菜单项的内容，分别为"添加""删除""查询"和"退出"。单击菜单键，运行结果如图 2.24 所示。

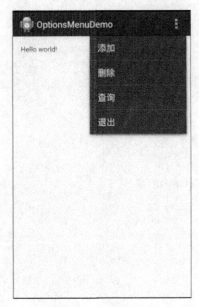

图 2.24　选项菜单

- 案例代码

Activity 代码：

```java
public class MainActivity extends Activity {
    @Override
    protected void onCreate(Bundle savedInstanceState) {
        super.onCreate(savedInstanceState);
        setContentView(R.layout.activity_main);
    }
//重写父类的onCreateOptionsMenu()方法
    public boolean onCreateOptionsMenu(Menu menu) {
        getMenuInflater().inflate(R.menu.menu_options, menu);   //加载菜单文件menu_options
        return true;
    }
```

```java
//重写父类的onOptionsItemSelected()方法
public boolean onOptionsItemSelected(MenuItem item) {
    switch(item.getItemId())     //判断点击的菜单项的id
    {
    case R.id.menu_add:
    {
        Log.i("Menu", "menu add");
        return true;
    }
    case R.id.menu_delete:
    {
        Log.i("Menu", "menu delete");
        return true;
    }
    case R.id.menu_query:
    {
        Log.i("Menu", "menu query");
        return true;
    }
    case R.id.menu_exit:
    {
        Log.i("Menu", "menu exit");
        return true;
    }
    default:
        return super.onOptionsItemSelected(item);
    }
}
```

menu_options.xml 文件：

```xml
<menu xmlns:android="http://schemas.android.com/apk/res/android" >
    <item
        android:id="@+id/menu_add"
        android:title="添加"/>
    <item
        android:id="@+id/menu_delete"
        android:title="删除"/>
```

```
    <item
        android:id="@+id/menu_query"
        android:title="查询"/>
    <item
        android:id="@+id/menu_exit"
        android:title="退出"/>
</menu>
```

- **案例分析**

在 Activity 中，重写了父类的 onCreateOptionsMenu(Menu menu)方法，在其中将自定义的菜单文件加入到 menu 中。然后重写了 onOptionsItemSelected(MenuItem item)方法，在其中调用了 MenuItem 的 getItemId()获取单击菜单项的 ID，根据不同的 ID 做对应的处理。

2.5.2 上下文菜单

上下文菜单和选项菜单的创建类似，也需要重写父类的相关方法，只不过创建上下文菜单重写父类的 onCreateContextMenu(ContextMenu menu, View v, ContextMenuInfo menuInfo)方法，单击上下文菜单中的某一项时，会回调 Activity 的 onContextItemSelected(MenuItem item)方法。另外，上下文菜单是长按某个 View 才会弹出来的菜单，所以还需要在代码中调用 registerForContextMenu(View view)将上下文菜单和 View 关联起来。下面我们通过一个实例说明一下上下文菜单的创建和使用。

案例 2.20 制作"添加""删除""查询"和"退出"上下文菜单

在 TextView 上长按，出现上下文菜单，运行结果如图 2.25 所示。

图 2.25 上下文菜单

- 案例代码

Activity 代码：

```java
public class MainActivity extends Activity {
    @Override
    protected void onCreate(Bundle savedInstanceState) {
        super.onCreate(savedInstanceState);
        setContentView(R.layout.activity_main);
        TextView txt = (TextView)findViewById(R.id.long_click);
        registerForContextMenu(txt);     //将View和上下文菜单关联
    }
    //重写父类的onCreateContextMenu()方法
    public void onCreateContextMenu(ContextMenu menu, View v,
            ContextMenuInfo menuInfo) {
        super.onCreateContextMenu(menu, v, menuInfo);
        getMenuInflater().inflate(R.menu.menu_context, menu);
    }
    //重写父类的onContextItemSelected()方法
    public boolean onContextItemSelected(MenuItem item) {
        switch(item.getItemId())
        {
        case R.id.menu_add:
        {
            Log.i("Menu", "menu add");
            return true;
        }
        case R.id.menu_delete:
        {
            Log.i("Menu", "menu delete");
            return true;
        }
        case R.id.menu_query:
        {
            Log.i("Menu", "menu query");
            return true;
        }
        case R.id.menu_exit:
        {
```

```
                    Log.i("Menu", "menu exit");
                    return true;
                }
            default:
                    return super.onContextItemSelected(item);
            }
        }
    }
```

- **案例分析**

在 Activity 中，重写了父类的 onCreateContextMenu(ContextMenu menu, View v, ContextMenuInfo menuInfo)方法，在其中将自定义的菜单文件加入到 menu 中。然后重写了 onContextItem Selected (MenuItem item)方法，其中的处理和选项菜单一致。另外，在 onCreate()方法中调用了 registerFor ContextMenu()方法，将上下文菜单和一个 TextView 进行关联。

2.6 常用资源类型

在之前的实例中，布局文件和代码中用到的字符串、颜色、尺寸全部都写成了固定值，这种编写方式会大大增加程序的开发和维护成本。例如对于一个应用来说，很多字体的大小和颜色都是相同的，每个布局文件在自己的 TextView 中都设置一个相同的颜色值和大小，如果其中一个 TextView 的属性不小心设置错误了，则会出现不统一的现象。另外，在程序后期的维护过程中，如果需要改变字体的大小和颜色，则需要对所有 TextView 的属性做更改。最好的实现方式是在一个地方定义好需要的颜色和尺寸，然后在所有需要使用的地方引用已经定义好的数值。如果需要改动，则只需要改动一个地方即可。Android 定义了资源的概念，所有的字符串常量、颜色值等都可以定义在资源中。另外，Android 中的资源还有助于实现国际化和不同设备的适配。

2.6.1 资源的类型和使用

Android 中的资源都放在项目的 res 目录下，res 目录包括多个子目录，对应不同的资源类型。Android 中支持的资源类型如表 2.9 所示。

表 2.9 Android 支持的资源类型

目录	资源类型
res/animator/	存放 xml 文件，定义属性动画
res/anim/	存放 xml 文件，定义补间动画
res/color/	存放 xml 文件，定义颜色状态列表
res/drawable/	存放图片或者 xml 文件，用于表示可绘制对象

续表

目录	资源类型
res/layout/	存放 xml 文件，定义页面布局
res/menu/	存放 xml 文件，定义菜单内容
res/raw/	以原始形式保存的任意文件。要以 IO 流的方式打开
res/values/	包含多种数值文件，相应的文件名如下。 arrays.xml：用于资源数组（类型化数组） colors.xml：颜色值 dimens.xml：尺寸值 strings.xml：字符串值 styles.xml：样式

创建 Android 项目之后，开发工具会自动在项目的 gen 目录下生成一个 R.java 的类文件，用于维护 Android 中每种类型资源的 ID。R.java 的内容示例如下：

```
/* AUTO-GENERATED FILE.  DO NOT MODIFY.
 *
 * This class was automatically generated by the
 * aapt tool from the resource data it found. It
 * should not be modified by hand.
 */
package com.demo.optionsmenu;

public final class R {
    public static final class attr {
    }
    public static final class dimen {
        public static final int activity_horizontal_margin=0x7f040000;
        public static final int activity_vertical_margin=0x7f040001;
    }
    public static final class drawable {
        public static final int ic_launcher=0x7f020000;
    }
    public static final class id {
        public static final int menu_add=0x7f080000;
        public static final int menu_delete=0x7f080001;
        public static final int menu_exit=0x7f080003;
        public static final int menu_query=0x7f080002;
```

```
    }
    public static final class layout {
        public static final int activity_main=0x7f030000;
    }
    public static final class menu {
        public static final int menu_options=0x7f070000;
    }
    public static final class string {
        public static final int action_settings=0x7f050001;
        public static final int app_name=0x7f050000;
        public static final int hello_world=0x7f050002;
    }
    public static final class style {
        public static final int AppTheme=0x7f060001;
    }
}
```

所以可以借助 R.java 文件访问资源，Android 中资源的使用有两种场景：在代码中访问和在 xml 文件中访问。

（1）在代码中访问

访问形式：R.<resource_type>.<resource_name>，其中 resource_type 代表资源的类型，resource_name 代表资源的名称。访问示例如下：

```
TextView txt = (TextView)findViewById(R.id.txt);
txt.setText(R.string.txt_name);
```

（2）在 xml 文件中访问

访问形式：@<resource_type>/<resource_name>，访问示例如下：

```xml
<TextView
    android:layout_width="fill_parent"
    android:layout_height="fill_parent"
    android:textColor="@color/blue"
    android:text="@string/txt_name" />
```

2.6.2 字符串、颜色、尺寸

字符串、颜色、尺寸的资源文件均在 res/values/目录下，三者都定义在 xml 文件中，根元素是<resouces>，每个子元素<string></string>定义一个字符串，每个子元素<color></color>定义一个颜色值，每个子元素<dimen></dimen>定义一个尺寸值。以字符串举例，定义的形式为<string name="XXX">XXX</string>，其中 name

字符串、颜色、尺寸资源

属性指定了该字符串的名称，便于引用，<string>和</string>之间就是字符串具体的内容。下面我们通过一个示例说明一下字符串、颜色、尺寸的具体定义和使用。

案例 2.21　字符串、颜色、尺寸的具体定义和使用

本例布局文件中定义两个 TextView 控件，分别引用不同的颜色值和尺寸值显示不同的文字，引用方式为@<resource_type>/<resource_name>，运行结果如图 2.26 所示。

图 2.26　字符串、颜色、尺寸资源

- 案例代码

strings.xml：

```
<?xml version="1.0" encoding="utf-8"?>
<resources>
    <string name="hello">Hello</string>
    <string name="world">World</string>
</resources>
```

colors.xml

dimens.xml：

```
<?xml version="1.0" encoding="utf-8"?>
<resources>
    <color name="red">#FF0000</color>
    <color name="green">#00FF00</color>
</resources>
<resources>
    <dimen name="text_little">20sp</dimen>
```

```xml
    <dimen name="text_normal">26sp</dimen>
</resources>
```

布局文件：

```xml
<LinearLayout xmlns:android="http://schemas.android.com/apk/res/android"
    xmlns:tools="http://schemas.android.com/tools"
    android:layout_width="match_parent"
    android:layout_height="match_parent"
    android:orientation="horizontal"
    >
    <TextView
        android:layout_width="wrap_content"
        android:layout_height="wrap_content"
        android:textColor="@color/red"    //引用颜色
        android:textSize="@dimen/text_little"    //引用尺寸
        android:text="@string/hello" />    //引用字符串
    <TextView
        android:layout_width="wrap_content"
        android:layout_height="wrap_content"
        android:textColor="@color/green"
        android:textSize="@dimen/text_normal"
        android:text="@string/world" />
</LinearLayout>
```

2.6.3 Drawable

Drawable 资源是指一些图片资源或者一些特定的可用于绘制的 xml 文件，本节我们将介绍常用的 3 种 Drawable 资源：图片资源、State List 和 Shape Drawable。

1. 图片资源

图片资源比较简单，只要将常用格式的图片放入到 /res/drawable-XXX 目录下，编译器会自动在 R.java 文件中生成对应的资源 ID。可以使用正常的资源访问方式使用图片。下面我们通过一个实例说明图片资源的使用方法。

> 案例 2.22　使用图片资源

本例布局文件比较简单，定义了一个 ImageView 控件，使用 @drawable/pic1 引用了一张图片用于显示，只需要将图片 pic1.png 放入 /res/drawable-XXX 目录下即可。运行结果如图 2.27 所示。

- **案例代码**

布局文件：

```xml
<RelativeLayout xmlns:android="http://schemas.android.com/apk/res/android"
```

```
    xmlns:tools="http://schemas.android.com/tools"
    android:layout_width="match_parent"
    android:layout_height="match_parent"
    >
    <ImageView
        android:layout_width="100dp"
        android:layout_height="100dp"
        android:scaleType="centerCrop"
        android:src="@drawable/pic1" />
</RelativeLayout>
```

图 2.27 图片资源

2. State List

一个控件可能会有不同的状态，例如获焦状态、被按下状态。State List 用于控制 View 在不同状态下的不同显示。定义 State List 的 xml 文件以<selector>作为根元素，中间可以有一个或多个<item>元素，用于标识不同的状态。表 2.10 所示的是<item>元素支持的属性。

表 2.10　State List 支持的属性

属性	属性说明
android:drawable	用于设置一个 Drawable 资源
android:state_pressed	设置是否处于按下状态
android:state_focused	设置是否处于获焦状态
android:state_selected	设置是否处于选中状态
android:state_checked	设置是否处于勾选状态

续表

属性	属性说明
android:state_enabled	设置是否处于可用状态
android:state_activated	设置是否处于活动状态
android:state_window_focused	设置窗口是否获得焦点

下面我们通过一个具体的实例说明一下 State List 的使用方法。

案例 2.23　使用 State List 制作按钮按下变色效果

为了做对比，本例布局文件中定义了两个一样的按钮，其中，android:textColor 属性均设置为了 @drawable/drawable_button，在 drawable_button.xml 文件中，设置了当按钮处于非按下状态时，文本颜色显示为红色，当按钮处于按下状态时，文本颜色显示为绿色，运行结果如图 2.28 所示。

图 2.28　State List

- **案例代码**

布局文件：

```
<LinearLayout xmlns:android="http://schemas.android.com/apk/res/android"
    xmlns:tools="http://schemas.android.com/tools"
    android:layout_width="match_parent"
    android:layout_height="match_parent"
    android:orientation="vertical" >
    <Button
        android:layout_width="wrap_content"
        android:layout_height="wrap_content"
```

```xml
        android:textSize="24sp"
        android:text="Click me"
        android:textColor="@drawable/drawable_button"
        />
    <Button
        android:layout_width="wrap_content"
        android:layout_height="wrap_content"
        android:textSize="24sp"
        android:text="Click me"
        android:textColor="@drawable/drawable_button"
        />
</LinearLayout>
```

drawable_button.xml：

```xml
<?xml version="1.0" encoding="utf-8"?>
<selector xmlns:android="http://schemas.android.com/apk/res/android" >
    <!-- 非按下状态时的颜色 -->
    <item
        android:state_pressed="false"
        android:color="#FF0000"/>
    <!-- 按下状态时的颜色 -->
    <item
        android:state_pressed="true"
        android:color="#00FF00"/>
</selector>
```

3. Shape Drawable

Shape Drawable 可以定义一个基本的几何图形，可以用来改变控件的外观显示。定义 Shape Drawable 的 xml 文件的根元素为<shape>，其中，可以使用 android:shape 属性设置定义的几何图形，取值为 rectangle、oval、line、ring，分别代表矩形、椭圆、水平线和环形。下面我们通过一个实例改变 EditText 的外观显示，具体说明一下 Shape Drawable 的使用方法。

案例 2.24 使用 Shape Drawable 制作圆角矩形的编辑框

为了做对比，本例布局文件中定义了两个相同的 EditText 组件，唯一的区别在于第 2 个 EditText 设置了 android:background 属性，属性值为@drawable/drawable_edit。在 drawable_edit.xml 文件中定义了一个矩形的 shape，使用<gradient>属性设置矩形的渐变色，使用<corners>属性设置矩形的 4 个角的弧度，使用<stroke>属性设置矩形线的宽度和颜色。运行结果如图 2.29 所示，第 2 个 EditText 是一个圆角矩形的编辑框。

图 2.29 Shape Drawable

- 案例代码

布局文件：

```xml
<LinearLayout xmlns:android="http://schemas.android.com/apk/res/android"
    xmlns:tools="http://schemas.android.com/tools"
    android:layout_width="match_parent"
    android:layout_height="match_parent"
    android:orientation="vertical" >
    <EditText
        android:layout_width="200dp"
        android:layout_height="wrap_content"
        android:textSize="24sp"
        android:hint="username" />
    <EditText
        android:layout_width="200dp"
        android:layout_height="wrap_content"
        android:textSize="24sp"
        android:hint="password"
        android:background="@drawable/drawable_edit" />    //设置drawable背景
</LinearLayout>
```

drawable_edit.xml

```xml
<?xml version="1.0" encoding="utf-8"?>
<shape xmlns:android="http://schemas.android.com/apk/res/android"
```

```
        android:shape="rectangle" >          //形状为矩形
        <gradient    //定义渐变色
            android:startColor="#999999"     //渐变起始颜色
            android:endColor="#FFFFFF"       //渐变结束颜色
            android:angle="0"/>    //渐变角度 0代表从左到右,90代表从上到下
        <corners     //定义角为圆角
            android:radius="8dp"/>
        <stroke    //定义线的颜色和宽度
            android:width="2dp"
            android:color="#0000FF"/>
</shape>
```

2.6.4 Style

Style 即样式,它定义了界面显示的风格。Style 可使用于一个单独的控件,也可以使用于一个 Activity 或者整个应用。使用 Style 的好处是可以减少代码的工作量,例如在 Android 应用中,许多地方的文本大小、颜色可能一致,所以每个 TextView 都要设置 android:textSize 和 android:textColor 属性,这时候就可以将这两个属性直接定义在 Style 中,每个 TextView 直接引用 Style 属性即可。Style 类似于 Word 中的样式,在 Word 中定义一个样式,设置其字体大小、间距等,每段文字只要使用这个样式,就具有该样式所定义的格式。下面我们通过一个具体的案例说明一下 Style 的使用。

案例 2.25　使用 Style 统一设置文字的大小和颜色

本例在布局文件中定义两个文本框,但是都不设置文本字体的大小和颜色,而是设置 style 属性,属性值为@style/TextStyle。在 style.xml 文件中,定义一个名为 TextStyle 的样式,统一设置文字的大小和颜色,运行结果如图 2.30 所示。

图 2.30　Style

- **案例代码**

布局文件：

```xml
<LinearLayout xmlns:android="http://schemas.android.com/apk/res/android"
    xmlns:tools="http://schemas.android.com/tools"
    android:layout_width="match_parent"
    android:layout_height="match_parent"
    android:orientation="vertical" >
    <TextView
        android:layout_width="wrap_content"
        android:layout_height="wrap_content"
        android:text="Hello"
        style="@style/TextStyle" />
    <TextView
        android:layout_width="wrap_content"
        android:layout_height="wrap_content"
        android:text="World"
        style="@style/TextStyle" />
</LinearLayout>
```

style.xml：

```xml
<resources xmlns:android="http://schemas.android.com/apk/res/android">
    <style name="TextStyle" >
        <item name="android:textSize">20sp</item>
        <item name="android:textColor">#FF0000</item>
    </style>
</resources>
```

2.6.5 国际化

Android 的资源除了可以提高代码的可编写性和维护性之外，还可以实现国际化。国际化是指同一个应用在不改变逻辑结构的情况下，在不同的语言环境中可以有不同的显示。例如将手机的语言环境切换成英文，许多应用会自动显示为英语。

在 Android 资源中实现国际化比较简单，只需要按照一定格式为不同的语言定义对应的资源文件夹，应用运行的时候会自动匹配加载最合适的文件。以字符串资源为例，实现国际化需要在 res 目录下创建对应语言 values 文件夹，values 文件夹的命名方式是 values-语言码-r 国家码，例如 values-zh-rCN 代表简体中文，其中，zh 代表中文，CN 代表中国大陆地区，类似的还有 values-en-rUS 代表美式英语。每个 values 文件夹下都有一个 strings.xml 文件，其中的字符串以不同的语言显示。下面我们通过一个实例说明一下字符串的国际化。

案例 2.26　制作同样的按钮在不同的语言环境下的显示效果

本例界面定义两个按钮，分别以@string 的方式引用字符串资源 btn_ok 和 btn_cancel，然后在 res 文件夹下建立简体中文和美式英文的 values 文件，为字符串 btn_ok 和 btn_cancel 提供不同语言下的显示内容，运行结果如图 2.31 所示。在不同的语言环境下，同样的按钮有不同的显示。

图 2.31　国际化

- 案例代码

布局文件：

```
<LinearLayout xmlns:android="http://schemas.android.com/apk/res/android"
    xmlns:tools="http://schemas.android.com/tools"
    android:layout_width="match_parent"
    android:layout_height="match_parent"
    android:orientation="vertical"
    android:padding="10dp" >
    <Button
        android:layout_width="50dp"
        android:layout_height="wrap_content"
        android:textSize="20sp"
        android:text="@string/btn_ok" />
    <Button
        android:layout_width="50dp"
        android:layout_height="wrap_content"
        android:textSize="20sp"
        android:text="@string/btn_cancel" />
</LinearLayout>
```

res/values-zh-rCN/strings.xml：

```
<?xml version="1.0" encoding="utf-8"?>
```

```xml
<resources>
    <string name="btn_ok">OK</string>
    <string name="btn_cancel">Cancel</string>
</resources>
```

res/values-en-rUS/strings.xml：

```xml
<?xml version="1.0" encoding="utf-8"?>
<resources>
    <string name="btn_ok">确认</string>
    <string name="btn_cancel">取消</string>
</resources>
```

2.7 事件处理和消息传递

在 Android 应用中，用户经常会在界面上进行各种操作，程序需要及时做出反馈。用户所做的操作称为一个事件，程序做出的反馈称为事件处理。Android 中的事件处理有两种：基于监听的事件处理和基于回调的事件处理，本节我们将分别介绍这两种事件处理的使用方式。另外，本节我们还将介绍一个重要的消息传递机制：Handler。

事件处理和消息传递

2.7.1 基于监听的事件处理

基于监听的事件处理方式我们在之前介绍 Android 基本组件中已经多次使用到，主要要设计 3 个对象。

（1）事件源：事件产生所在的组件，例如单击一个按钮，按钮就是事件源。

（2）事件类型：产生的事件类型，如单击事件、长按事件、触摸事件等。

（3）事件监听器：被动地监听组件上产生的事件，并做出相应处理。

使用基于监听的事件处理时，需要对组件调用相应的 setListener()方法设置事件监听器，例如调用 setOnClickListener()方法监听单击事件，调用 setOnLongClickListener()方法监听长按事件，并重写其中的回调方法做自定义的处理。用户对界面进行操作时，就会触发事件监听器，事件监听器会调用对应的方法做处理，例如对于单击事件，系统会调用 OnClickListener 监听器的 onClick()方法做处理。

基于监听的事件处理方式之前已经有多个实例，如 2.3.2 节的按钮组件，这里我们就不再给出实例说明了。

2.7.2 基于回调的事件处理

基于监听的事件处理是将事件源和事件监听器分开，而基于回调的事件处理则不需要为组件设置事件监听器，当系统检测到有用户操作时，会直接回调组件中特定的方法做处理。因此，基于回调的事件处理需要定义一个类继承自需要的组件，并重写其中的特定方法以实现自定义的处理。下面我们仍以按钮组件为例,说明一下基于回调事件的处理方式。

案例 2.27　基于回调事件的处理

本例布局文件中使用一个自定义的组件 CustomButton，CustomButton 比较简单，继承自 Button，所以和 Button 有一样的功能，只不过重写了 Button 的 onKeyUp()方法，在其中实现自定义的处理。单击按钮之后，这里会打印一条日志。

- 案例代码

xml 布局文件：

```xml
<RelativeLayout xmlns:android="http://schemas.android.com/apk/res/android"
    xmlns:tools="http://schemas.android.com/tools"
    android:layout_width="match_parent"
    android:layout_height="match_parent"
    >
    <com.demo.view.CustomButton    //添加一个自定义的按钮组件
        android:layout_width="wrap_content"
        android:layout_height="wrap_content"
        android:text="自定义按钮"
        android:textSize="20sp" />
</RelativeLayout>
```

CustomButton 代码：

```java
public class CustomButton extends Button {    //定义一个自定义的按钮组件
    public CustomButton(Context context, AttributeSet attrs) {
        super(context, attrs);
    }
    //重写父类的onKeyUp()事件实现自定义的处理
    public boolean onKeyUp(int keyCode, KeyEvent event) {
        super.onKeyUp(keyCode, event);
        Log.i("button", "Button onKeyUp Event");
        return true;    //返回true,事件会被该组件消耗掉，否则会继续向上传递
    }
}
```

在 Activity 中，也有大量的事件回调方法，用于在整个 Activity 页面对事件进行处理，开发者可以实现其中的方法做自定义的处理。

2.7.3　Handler 消息传递

在 Android 中，UI 的刷新只能在主线程中进行，开发者新启动的子线程中不能做 UI 相关的操作，这种情况可以通过 Handler 类进行处理。Handler 的处理方式是通过一系列的 post 和 send 方法发送一

条消息到消息队列中,系统会从消息队列中取出消息并执行对应的任务。Handler 还可以延时发送消息,所以可用于延时操作。常见的消息发送方法如表 2.11 所示。

表 2.11　Handler 的消息发送方法

方法名	方法说明
post(Runnable r)	将任务加入到消息队列中
postAtTime(Runnable r, long uptimeMillis)	在指定时间将任务加入到消息队列中
postDelayed(Runnable r, long delayMillis)	延迟一定时间后将任务加入到消息队列中
removeCallbacks(Runnable r)	将任务从消息队列中移除
sendEmptyMessage(int what)	发送消息 ID 为 what 的消息
sendMessage(Message msg)	发送消息体为 msg 的消息

下面我们通过一个实例说明一下 Handler 的使用方法。

案例 2.28　基于回调事件的处理

本例在 Activity 中使用 Timer 类启动一个定时任务,每隔一秒生成一个随机数,并通过 Handler 的 sendMessage()方法发送一条消息。然后自定义一个类继承自 Handler,重写父类的 handleMessage() 方法对消息进行处理。

- 案例代码

Activity 代码:

```java
public class MainActivity extends Activity {
    private final int MSG_ID = 1001;    //定义一个消息ID
    private TextView txt;
    private MyHandler mHandler = new MyHandler();    //定义一个Handler对象
    protected void onCreate(Bundle savedInstanceState) {
        super.onCreate(savedInstanceState);
        setContentView(R.layout.activity_main);
        txt = (TextView)findViewById(R.id.txt);
        new Timer().schedule(new TimerTask(){    //使用Timer启动一个定时任务
            public void run() {
                int num = new Random().nextInt(1000);    //生成一个随机数
                Message msg = new Message();    //定义一个消息体
                msg.what = MSG_ID;    //设置消息ID
                msg.obj = num;    //设置消息内容
                mHandler.sendMessage(msg);    //使用Handler发送消息
            }
        }, 1000, 1000);    //每隔一秒执行一次
```

```
        }
    class MyHandler extends Handler      //自定义一个类继承自Handler类
    {
        public void handleMessage(Message msg) {    //重写父类的handleMessage()方法
            switch(msg.what)     //判断消息ID
            {
            case MSG_ID:
            {
                int num = (Integer)msg.obj;
                txt.setText("The num is: " + num);    //改变TextView的显示
                break;
            }
            default:
                super.handleMessage(msg);
                break;
            }
        }
    }
}
```

2.8 本章小结

本章我们主要介绍了 Android 界面开发的一些基础知识，对于 Android 应用来说，界面的美观性和友好性十分影响用户体验，所以掌握 Android 的界面开发非常重要。在本章我们首先介绍了视图和视图容器等基本概念，然后介绍了一些常用的布局方式。Android 提供了各种各样的控件供开发页面使用，本章我们介绍了用于显示文本的文本框和编辑框，用于显示图片的图片控件，还有一些用于操作的按钮和菜单，以及用于提示的对话框。最后介绍了一些常用的资源类型，Android 中的事件处理和模块与模块之间的消息传递。

Android 页面的布局方式和控件使用常常根据应用的具体场景做不同的选择，在后续的 3～8 章中我们将通过一些练习继续熟悉这些组件的使用。

第3章

Activity

■ Android 应用有 4 大组件，分别是 Activity（活动）、Service（服务）、Content Provider（内容提供者）和 Broadcast Receiver（广播接收器）。其中，Activity 是最常用的组件，它提供一个屏幕，用户可以用来交互，完成某项任务。在第 2 章介绍布局的时候，我们已经使用过 Activity，用它绘制用户界面并响应用户的操作。

一个程序通常会包括多个 Activity，有一个 Activity 会被声明为主 Activity，用户进入应用时第一个显示的就是主 Activity，然后可以根据不同的操作跳转到其他 Activity。本章我们将会介绍 Activity 开发的相关知识。

3.1 Activity 的使用

定义一个自己的 Activity 需要继承基类 Activity，并且需要实现基类中的若干方法，其中必须实现的方法是 onCreate()方法，Activity 刚启动的时候会回调该方法，所以可以在其中做一些初始化操作，其中最重要的是需要调用 Activity 的 setContentView(View view)方法设置布局文件。下面我们通过一个示例代码说明如何创建 Activity。

Activity 的使用和跳转

创建 Activity 示例：

```
public class MainActivity extends Activity {    继承基类Activity

    protected void onCreate(Bundle savedInstanceState) {    //重写父类的onCreate方法
        super.onCreate(savedInstanceState);
        setContentView(R.layout.act_main);    //调用setContentView设置布局文件
    }
}
```

在创建完成 Activity 之后，还需要在 AndroidManifest.xml 文件中对 Activity 进行配置，否则系统找不到定义的 Activity。Activity 的配置方法是在 manifest 文件的<application/>元素中增加一个<activity/>元素，代码示例如下，其中 android:name 属性指定了 MainActivity 的位置。

```
<manifest ... >
  <application ... >
    <activity android:name="com.example.activity.MainActivity" />
    ...
  </application ... >
  ...
</manifest >
```

除了 android:name 属性，在 Activity 中还可以配置一些其他属性，例如可以通过 android:label 指定 Activity 的书签，通过 android:icon 指定 Activity 的图标，通过 android:theme 指定 Activity 的样式等。

另外，Activity 通常还需要配置一个或多个<intent-filter></intent-filter>属性，用于指定该 Activity 可响应的 Intent。在第 4 章中，我们会详细介绍关于 Intent 和 intent-filter 属性的知识。

3.2 Activity 之间的跳转

应用程序一般都由多个界面组成，用户不同的操作会触发显示不同的界面，这就要求 Activity 之间可以实现跳转和数据交换。启动一个 Activity 有如下两种方法。

（1）startActivity(Intent intent)：启动其他 Activity。

（2）startActivityForResult(Intent intent, int requestCode)：该方法不仅可以启动其他 Activity，还可以接收其他 Activity 的返回结果，requestCode 标识请求的来源，可以自定义。

上述两个方法的参数中均含有 Intent 类型的参数，Intent 可以理解为"意图"，它可以设置将要启动的 Activity，也可以携带部分数据。Intent 重载了一系列 putExtra 方法用于传递数据，同时提供了一系列的 get 方法用于取出携带的数据。下面我们通过两个具体的案例介绍一下 Activity 的两种启动方式。

案例 3.1　用 startActivity 方法实现跳转

本实例定义两个界面，在第 1 个界面中定义一个 Button，用于单击进行跳转，第 2 个界面接收第 1 个界面传递的数据并将其显示出来，运行结果如图 3.1 所示，左图界面是 ActivityA，单击按钮后跳转到右图界面。

图 3.1　Activity 跳转方式一

- 案例代码

ActivityA 代码：

```java
public class ActivityA extends Activity {

    //实现基类Activity的onCreate方法
    protected void onCreate(Bundle savedInstanceState) {
        super.onCreate(savedInstanceState);
        setContentView(R.layout.act_a);      //调用setContentView函数设置布局文件
        initWidget();
    }


    private void initWidget()
    {
```

```java
        Button btn = (Button)findViewById(R.id.btn_click);    //获取"跳转"按钮
        btn.setOnClickListener(new OnClickListener(){    //为按钮设置单击监听事件
            public void onClick(View arg0) {
                //定义一个Intent对象,设置要跳转到的Activity
                Intent intent = new Intent(ActivityA.this, ActivityB.class);
                intent.putExtra("key", "this is a message");    //携带数据
                startActivity(intent);    //调用startActivity函数进行跳转
            }
        });
    }
```

ActivityB 代码:

```java
public class ActivityB extends Activity {
    protected void onCreate(Bundle savedInstanceState) {
        super.onCreate(savedInstanceState);
        setContentView(R.layout.act_b);
        Intent intent = getIntent();    //调用getIntent方法获取Intent对象
        //根据键值获取上个页面传递过来的数据
        String strValue = intent.getStringExtra("key");
        TextView txtVew = (TextView)findViewById(R.id.txt_content);
        txtVew.setText(strValue);    //将数据显示出来
    }
```

Activity 配置文件:

```xml
<application
    android:allowBackup="true"
    android:icon="@drawable/ic_launcher"
    android:label="@string/app_name"
    android:theme="@style/AppTheme" >
    <activity
        android:name="com.demo.activity.ActivityA"
        android:label="@string/app_name" >
        <intent-filter>
            <action android:name="android.intent.action.MAIN" />
            <category android:name="android.intent.category.LAUNCHER" />
        </intent-filter>
    </activity>
    <activity android:name="com.demo.activity.ActivityB"/>
```

		</application>

案例 3.2　用 startActivityForResult 方法实现登录效果

本例运行结果如图 3.2 所示，左图界面是 MainActivity 的起始状态，只有一个按钮，单击之后会跳转到 LoginActivity，即中间的图所示的界面。在 LoginActiviy 界面输入用户名和密码之后单击"登录"，又会返回到 MainActivity，并显示用户输入的信息。

图 3.2　Activity 的跳转方式二

- **案例代码**

MainActivity 代码：

```java
public class MainActivity extends Activity {
    @Override
    protected void onCreate(Bundle savedInstanceState) {
        super.onCreate(savedInstanceState);
        setContentView(R.layout.act_main);
        initWidget();
    }

    private void initWidget() {
        Button btn = (Button) findViewById(R.id.btn_login);    //获取按钮控件
        btn.setOnClickListener(new OnClickListener() {
            public void onClick(View v) {
                Intent intent = new Intent(MainActivity.this,
                        LoginActivity.class);    //定义Intent变量，设置将要跳转的Activity
                startActivityForResult(intent, 1);    //调用ActivityForResult启动另外一个Activity
            }
        });
```

```java
        }

        //重写父类的onActivityResult方法,接收其他界面的返回结果
        @Override
        protected void onActivityResult(int requestCode, int resultCode, Intent data) {
            Log.e("result", "requestCode:" + requestCode + "; resultCode:" + resultCode + "; data:" + data);
            if (1 != requestCode || RESULT_OK != resultCode || null == data) {
                return;
            }
            String strUsername = data.getStringExtra("username");   //取出用户名
            String strPassword = data.getStringExtra("password");   //取出密码
            showResult(strUsername, strPassword);    //显示结果
        }

        private void showResult(String username, String password) {
            Button btn = (Button) findViewById(R.id.btn_login);
            btn.setVisibility(View.GONE);
            LinearLayout lly = (LinearLayout) findViewById(R.id.user_info);
            lly.setVisibility(View.VISIBLE);

            TextView txtUsername = (TextView) findViewById(R.id.txt_username);
            TextView txtPassword = (TextView) findViewById(R.id.txt_password);
            txtUsername.setText("用户输入的用户名是:" + username);
            txtPassword.setText("用户输入的密码是:" + password);
        }
    }
```

LoginActivity 代码:

```java
    public class LoginActivity extends Activity {

        protected void onCreate(Bundle savedInstanceState) {
            super.onCreate(savedInstanceState);
            setContentView(R.layout.act_login);
            initWidget();
        }

        private void initWidget() {
```

```
            Button btn = (Button) findViewById(R.id.btn_commit);
            btn.setOnClickListener(new OnClickListener() {
                public void onClick(View v) {
                    EditText editUserName = (EditText) findViewById(R.id.edit_username);    //获取输入用户名的控件
                    EditText editPassword = (EditText) findViewById(R.id.edit_password);    //获取输入密码的控件
                    Intent data = new Intent();    //定义一个Intent用于传递数据
                    data.putExtra("username", editUserName.getText().toString());    //设置用户名
                    data.putExtra("password", editPassword.getText().toString());    //设置密码
                    setResult(RESULT_OK, data);    //设置返回结果
                    LoginActivity.this.finish();    //把自己finish掉
                }
            });
        }
```

- **案例分析**

本例在 MainActivity 中定义了一个按钮,按钮被单击后会启动 LoginActivity,启动的方式是 startActivityForResult(),并在 MainActivity 中重写例如父类的 onActivityResult()方法,用于接收从 LoginActivity 返回的数据。在 LoginActivity 中定义了两个输入框用于输入用户名和密码,当按钮被单击后,获取输入框的内容,并通过 setResult()方法将数据返回给 MainActivity 显示。

3.3 Activity 的生命周期

Activity 的生命周期

当 Android 应用在系统中运行时,每个 Activity 都有它的生命周期。如果要想编写出一个健壮、灵活的程序,掌握其生命周期至关重要。Activity 从创建到销毁,可以分为如下 3 种状态。

(1)活动状态:当前 Activity 处于前台,可以获得焦点,可以被用户看见并响应用户的操作。

(2)暂停状态:当前 Activity 依然被用户可见,但是不能获得焦点,其他 Activity 处于前台。一个处于暂停状态的 Activity 仍然处于内存中,但是在系统内存较低的时候可能会被回收掉。

(3)停止状态:当前 Activity 不再可见,完全处于后台。当其他地方有内存需要时,该 Activity 会被回收。

Activity 的每一种状态都会有相对应的回调方法,图 3.3 显示了 Activity 的生命周期和对应的回调方法。

图 3.3 显示了一个 Activity 的完整的生命周期,当 Activity 被创建时,系统会回调 onCreate()方法,当 Activity 被销毁时,会回调 onDestroy()方法。onStart()和 onStop()包含了 Activity 的可见周期,当

Activity 开始可见的时候，会回调 onStart()方法，当 Activity 变为不可见的状态时，会回调 onStop()方法。onResume()和 onPause()包含了 Activity 处于前台的周期，当 Activity 可以获得焦点，可以和用户交互时，会回调 onResume()方法，当 Activity 处于暂停状态时，会回调 onPause()方法。开发者可以在不同的回调方法中做自己的操作，例如在 onCreate()方法中做一些初始化的工作，在 onDestroy()方法中进行一些资源的释放。下面我们通过一个实例说明一下 Activity 回调方法执行的时机。

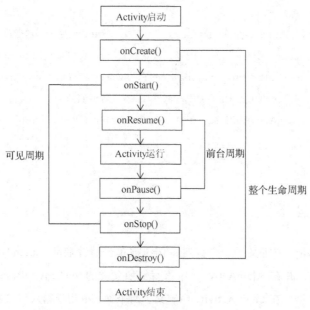

图 3.3　Activity 的生命周期和对应的回调方法

Activity 代码：

```
public class MainActivity extends Activity implements OnClickListener {

    private final String TAG = "life";

    protected void onCreate(Bundle savedInstanceState) {
        super.onCreate(savedInstanceState);
        Log.e(TAG, "=======onCreate=======");
        setContentView(R.layout.activity_main);
        initWidget();
    }

    @Override
    protected void onResume() {
        super.onResume();
        Log.e(TAG, "=======onResume=======");
```

```java
    }

    @Override
    protected void onPause() {
        Log.e(TAG, "======onPause======");
        super.onPause();
    }

    @Override
    protected void onStart() {
        super.onStart();
        Log.e(TAG, "======onStart======");
    }

    @Override
    protected void onStop() {
        Log.e(TAG, "======onStop======");
        super.onStop();
    }

    @Override
    protected void onDestroy() {
        Log.e(TAG, "======onDestroy======");
        super.onDestroy();
    }

    private void initWidget()
    {
        Button btn1 = (Button)findViewById(R.id.btn1);
        Button btn2 = (Button)findViewById(R.id.btn2);
        btn1.setOnClickListener(this);
        btn2.setOnClickListener(this);
    }

    @Override
    public void onClick(View v) {
        switch(v.getId())
```

```
            {
                case R.id.btn1:
                {
                    displayDialog();
                    break;
                }
                case R.id.btn2:
                {
                    MainActivity.this.finish();
                    break;
                }
                default:
                    break;
            }
```

在 MainActivity 中重写了生命周期的每个方法，并打印了相关的日志信息，当 Activity 刚启动的时候，日志打印信息如图 3.4 所示，可以看到 onCreate()、onStart()、onResume()方法被顺序调用。

L...	Time	PID	TID	Application	Tag	Text
E	05-01 11:40:47.626	654	654	com.demo.activity	life	=======onCreate=======
E	05-01 11:40:47.936	654	654	com.demo.activity	life	=======onStart=======
E	05-01 11:40:47.936	654	654	com.demo.activity	life	=======onResume=======

图 3.4 Activity 创建

单击手机上的 Home 键，Activity 会处于后台，但是并没有被销毁，日志打印信息如图 3.5 所示，onPause()、onStop()方法被顺序调用。

L...	Time	PID	TID	Application	Tag	Text
E	05-01 15:14:40.520	559	559	com.demo.activity	life	=======onPause=======
E	05-01 15:14:43.409	559	559	com.demo.activity	life	=======onStop=======

图 3.5 Activity 处于后台

当 Activity 处于后台时，此时重新单击该应用程序图标，让 Activity 重新显示在前台，日志打印信息如图 3.6 所示，onStart()、onResume()方法被顺序调用。

L...	Time	PID	TID	Application	Tag	Text
E	05-01 15:15:40.868	559	559	com.demo.activity	life	=======onStart=======
E	05-01 15:15:40.868	559	559	com.demo.activity	life	=======onResume=======

图 3.6 Activity 重新处于前台

单击界面上的"退出"按钮，执行 finish()方法，Activity 完全被销毁，日志打印信息如图 3.7 所示，onPause()、onStop()、onDestroy()方法被顺序调用。

图 3.7　Activity 完全被销毁

3.4　Activity 的启动模式

一个应用通常包括多个 Activity，每个 Activity 均有自己特定的操作，并且可以启动其他 Activity。例如一个邮箱应用，可能会有一个 Activity 显示邮件列表，用户单击其中一个邮件之后，会启动打开一个新的 Activity 查看邮件的具体内容。本节我们将介绍系统如何排列一个应用中的多个 Activity。

Activity 的启动模式

在 Android 中，每启动一个应用时，系统都会为该应用创建一个任务栈，并且将该应用的主页 Activity 压入堆栈底部。如果在当前 Activity 中打开一个新的 Activity，则系统会保存之前的 Activity 的状态，将新打开的 Activity 压入堆栈的顶部，并且获取焦点。当用户单击"返回"键之后，当前 Activity 会从堆栈顶部移除，之前的 Activity 恢复状态并正常运行。如果堆栈中已经不存在 Activity，则系统会将应用的任务栈回收掉。这是系统对于 Activity 的默认处理方式，也适用于大多数应用。但是有些时候，我们可能想要一些不同的处理方式。例如，当我们多次启动同一个 Activity 时，系统仍然会创建多个 Activity 对象并将其保存在堆栈中。对此，Android 提供了启动模式使用户能够修改这种默认的行为，可以通过两种方式指定 Activity 的启动方式：在 AndroidManifest.xml 中配置或在 Intent 中设置。

1. 在 AndroidManifest.xml 中配置

在配置中使用<activity>的 launchMode 属性指定当前 Activity 的启动方式，示例如下：

```
<activity
        android:name="com.demo.ActivityA"
        android:launchMode="standard">
    <intent-filter >
        <action android:name="android.intent.demo"/>
    </intent-filter>
</activity>
```

2. 在 Intent 中设置

通过 Intent 的 addFlags()方法指定启动 Activity 的方式，示例如下：

```
Intent intent = new Intent();
intent.setAction("android.intent.demo");
intent.addFlags(Intent.FLAG_ACTIVITY_NEW_TASK);
startActivity(intent);
```

Android 提供的启动模式有 4 种，分别是 standard、singleTop、singleTask、singleInstance。

（1）standard：默认的启动模式，每启动一个 Activity，都新建一个 Activity 对象并压入当前任务栈中。

（2）singleTop：系统启动一个 Activity，会判断当前待启动的 Activity 和栈顶的 Activity 是否一致，如果是同一个 Activity，则不新建当前 Activity 的对象，而是回调栈顶 Activity 对象的 onNewIntent() 方法。例如，当前任务栈中存在 ABCD 4 个 Activity，此时启动 Activity D，如果是 standard 模式，则会新建 D 的对象并压入栈中，则栈中的 Activity 为 ABCDD。如果是 singleTop 模式，则不新建对象，直接回调 D 的 onNewIntent() 方法，栈中的 Activity 仍为 ABCD。

（3）singleTask：新启动的 Activity 如果在当前任务栈中已经存在，则不新建对象，直接回调栈中已存在对象的 onNewIntent() 方法。该启动模式和 singleTop 类似，但是不要求新启动的 Activity 在栈顶存在，只要在栈中即可。

（4）singleInstance：每启动一个应用，系统都会为该应用建立一个任务栈。singIeInstance 启动模式要求 Activity 只能单独地位于一个任务栈中，即对象在所有的任务栈范围内都只存在一份。

3.5 本章小结

作为 Android 的四大组件之一，Activity 主要用于显示界面。在本章我们首先介绍了 Activity 的基本使用和跳转，然后介绍了 Activity 的整个生命周期，这样有助于读者更好地了解 Activity 从创建到消亡的整个过程。最后我们介绍了 Activity 的 4 种启动模式以及 Activity 在系统中的存储方式。

3.6 小练习

下面我们通过一个案例来练习 Activity 的具体使用，同时也熟悉一下第 2 章介绍的一些组件。

本练习旨在实现一个简易的图片浏览器，初始页面如图 3.8 所示，首先显示图片的分类和缩略图，单击图片的分类之后，可以预览该分类下的图片，如图 3.9 所示。单击网格中的图片可以显示对应的大图，如图 3.10 所示。

图 3.8　图片分类

图 3.9　图片缩略图

图 3.10　预览大图

下面我们来介绍具体的实现过程，首先是首页的 Activity，代码如下：

```
public class MainActivity extends Activity {
    @Override
```

```java
        protected void onCreate(Bundle savedInstanceState) {
            super.onCreate(savedInstanceState);
            setContentView(R.layout.activity_main);
            //创建Adapter
            ListViewAdapter adapter = new ListViewAdapter(this);

            ListView listView = (ListView)findViewById(R.id.lvew);
            listView.setAdapter(adapter);
            //设置单击事件
            listView.setOnItemClickListener(new OnItemClickListener() {
                @Override
                public void onItemClick(AdapterView<?> parent, View view, int pos,
                    long id) {
                    //跳转到其他Activity，并携带参数
                    Intent intent = new Intent(MainActivity.this, PreviewActivity.class);
                    intent.putExtra("startImgIndex", pos + 1);
                    intent.putExtra("endImgIndex", (pos + 1) * 9);
                    startActivity(intent);
                }
            });
        }
    }
```

从代码中可以看到，MainActivity 继承自 Activity，加载了一个 listview，listview 设置的 Adapter 为自定义的 ListViewAdapter，并且为 listview 的 item 设置了单击的监听事件，如果 item 被单击，会跳转到 PreviewActivity，并携带该分类下起始图片的序号和该分类的图片总数。其中，自定义的 ListViewAdapter 和 PreviewActivty 代码如下。

ListViewAdapter 代码：

```java
public class ListViewAdapter extends BaseAdapter{
    private List<ImageInfo> imgs;
    private LayoutInflater flater;

    public ListViewAdapter(Context context)
    {
        flater = LayoutInflater.from(context);
        imgs = new ArrayList<ImageInfo>();
        //从drawable中加载显示图片
```

```java
        imgs.add(new ImageInfo(R.drawable.demo1, "camera", 9));
        imgs.add(new ImageInfo(R.drawable.demo10, "wexin", 2));
    }

    public int getCount() {
        return (null == imgs || imgs.isEmpty()) ? 0 : imgs.size();
    }


    @Override
    public Object getItem(int pos) {
        if(null != imgs && pos < imgs.size())
        {
            return imgs.get(pos);
        }
        return null;
    }


    @Override
    public long getItemId(int position) {
        return 0;
    }


    @Override
    public View getView(int position, View convertView, ViewGroup parent) {
        if(null == convertView)
        {
            //为每一个子项加载布局
            convertView = flater.inflate(R.layout.view_list_item, null);
        }
        //获取子item布局文件中的控件
        ImageView imgView = (ImageView)convertView.findViewById(R.id.lvew_img);

        TextView txtTitle = (TextView)convertView.findViewById(R.id.lvew_title);
        TextView txtNumber = (TextView)convertView.findViewById(R.id.lvew_number);


        ImageInfo imgInfo = (ImageInfo)getItem(position);        //根据position获取数据
        if(null != imgInfo)
```

```java
            {
                txtTitle.setText(imgInfo.getTitle());
                txtNumber.setText(String.valueOf(imgInfo.getNumber()) + " 张照片");
                imgView.setImageResource(imgInfo.getImageId());
            }
            return convertView;
        }
    }
```

PreviewActivity 代码:

```java
public class PreviewActivity extends Activity{
    @Override
    protected void onCreate(Bundle savedInstanceState) {
        super.onCreate(savedInstanceState);
        setContentView(R.layout.activity_preview);

        //获取上一个Activity携带的参数
        Intent intent = getIntent();
        int startImgIndex = intent.getIntExtra("startImgIndex", 1);
        int endImgIndex = intent.getIntExtra("endImgIndex", 9);

        final GridViewAdapter adapter = new GridViewAdapter(this,startImgIndex,endImgIndex);
        //获取gridView
        GridView gridView = (GridView)findViewById(R.id.grdView);
        gridView.setAdapter(adapter);
        gridView.setOnItemClickListener(new OnItemClickListener() {

            @Override
            public void onItemClick(AdapterView<?> parent, View view, int pos,
                    long id) {
                //单击item跳转到大图页面,并传递图片的resource id
                Intent intent = new Intent(PreviewActivity.this, BigImgActivity.class);
                intent.putExtra("imgId", Integer.valueOf(adapter.getItem(pos). toString()));
                startActivity(intent);
            }
        });
    }
}
```

PreviewActivity 主要是通过 gridView 实现图片的网格显示，在 onCreate()方法中，通过 getIntent()方法获取该分类下的起始图片和图片总数，并传递到自定义的 Adapter 中用于显示。同样地，对 gridView 的 item 增加了单击事件，跳转到 BigImgActivity 显示大图，同时传递被单击图片的 resource id，用于大图的显示。GridViewAdapter 和 BigImgActivity 的代码如下。

GridViewAdapter 代码：

```java
public class GridViewAdapter extends BaseAdapter{

    private int[] allImgs = {R.drawable.demo1,R.drawable.demo2,R.drawable.demo3,R.drawable.demo4,R.drawable.demo5,R.drawable.demo6,R.drawable.demo7,R.drawable.demo8,R.drawable.demo9,R.drawable.demo10,R.drawable.demo11};

    private List<Integer> imgs;
    private LayoutInflater flater;
    public GridViewAdapter(Context context, int startImgIndex, int endImgIndex)
    {
        flater = LayoutInflater.from(context);
        imgs = new ArrayList<Integer>();
        //获取需要展示的图片集合
        for(int i = startImgIndex - 1; i < endImgIndex; ++i)
        {
            if(i > allImgs.length - 1)
            {
                break;
            }
            imgs.add(allImgs[i]);
        }
    }
    @Override
    public int getCount() {
        return (null == imgs || imgs.isEmpty()) ? 0 : imgs.size();
    }
    @Override
    public Object getItem(int pos) {
        if(null != imgs && pos < imgs.size())
        {
            return imgs.get(pos);
```

```java
        }
        return null;
    }
    @Override
    public long getItemId(int position) {
        return 0;
    }
    @Override
    public View getView(int position, View convertView, ViewGroup parent) {
        if(null == convertView)
        {
            //为每一个子项加载布局
            convertView = flater.inflate(R.layout.view_grid_item, null);
        }
        //获取子item布局文件中的控件
        ImageView imgView = (ImageView)convertView.findViewById(R.id.gvew_img);
        imgView.setImageResource(imgs.get(position));
        return convertView;
    }
}
```

BigImgActivity 代码：

```java
public class BigImgActivity extends Activity{
    @Override
    protected void onCreate(Bundle savedInstanceState) {
        super.onCreate(savedInstanceState);
        setContentView(R.layout.activity_big_img);
        Intent intent = getIntent(); //获取需要显示的图片id
        int imgId = intent.getIntExtra("imgId", R.drawable.demo1);
        ImageView imgView = (ImageView)findViewById(R.id.big_img);
        imgView.setImageResource(imgId);
    }
}
```

BigImgActivity 在 onCreate()方法中通过 getIntent()方法获取需要显示的图片 id，然后设置到 ImageView 控件中进行大图的显示。

第4章

Intent和BroadCastReceiver

■ 在第 3 章介绍 Activity 时，我们多次使用到了 Intent。Intent 是一个用于消息传递的对象，Android 系统可以根据 Intent 设置的内容去调用不同的组件，同时可以携带一些必要的数据。本章我们将详细介绍 Intent 的使用，另外还将介绍 Android 应用的另一个非常重要的应用——广播接收者。

4.1　Intent 和 intent-filter 配置

Intent 可以启动不同的组件，基本的使用场景有以下 3 种。

（1）将 Intent 对象作为参数传递给 startActivity()或 startActivityForResult()方法，启动一个 Activity。

（2）将 Intent 对象作为参数传递给 startService()或 bindService()，启动一个 Service。

（3）将 Intent 对象作为参数传递给 sendBroadcast()、sendOrderedBroadcast()或 sendSticky Broadcast()，发送一个广播。

关于 Service 的内容我们将会在第 7 章中介绍，此处暂不展开说明。下面我们来详细介绍一下 Intent 相关的知识。

Intent 可以分成两种类型，分别是显式 Intent 和隐式 Intent。

（1）显式 Intent：在定义 Intent 对象时根据类名明确指定所要启动的组件，这种方式通常用于应用程序的内部，例如从一个 Activity 界面跳转到另外一个界面，代码示例如下：

//Intent明确指定了要跳转到LoginActivity
Intent intent = new Intent(MainActivity.this, LoginActivity.class);
startActivity(intent);

（2）隐式 Intent：在定义 Intent 对象时，没有明确指定所要启动的组件，只是设置了一些过滤条件，Android 系统根据特定的匹配过滤规则，找到并启动相应的组件，代码示例如下：

Intent intent = new Intent();
//Intent并没有指定要启动哪个组件，只是设置了一个过滤规则
intent.setAction("com.action.show");
startActivity(intent);

可以看出，隐式 Intent 只是表明了"想要做什么"，它并不关心谁去"执行"它想要的这个操作，Android 系统会去匹配并启动合适的组件。

创建隐式 Intent 之后，系统会将 Intent 的内容与设备上其他应用的 AndroidManifest.xml 文件中声明的 intent-filter 进行比较，从而找到所要启动的组件。在第 3 章介绍 Activity 的配置时我们提到过，一个 Activity 组件可以声明多个 intent-filter，它定义了该组件可以接受的启动规则。一个 intent-filter 可以设置 action、data、category 属性，Android 系统也是依次通过这 3 个属性进行匹配判断的。下面我们将分别介绍这 3 个属性。

1. action 属性

action 属性是一个字符串，在定义隐式 Intent 时，如果通过 setAction()方法指定了 action，系统会查找设备中应用的 AndroidManifest 的文件，如果有组件在 intent-filter 属性里配置了相同的 action，则匹配成功，系统会启动该组件。下面我们通过一个实例说明一下 action 的匹配。

案例 4.1　通过 action 启动 activity

本例运行结果如图 4.1 所示，在用户单击 MainActivity 中的按钮之后，系统会在各个应用中的 AndroidManifest.xml 文件中查找 action 属性。匹配到 ShowActivity 配置的 action 属性一致，则启动 ShowActivity。

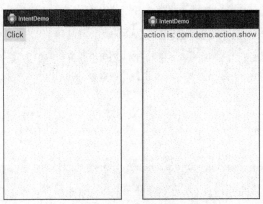

图 4.1　action 跳转的运行结果

- **案例代码**

MainActivity 代码：

```
public class MainActivity extends Activity {
    protected void onCreate(Bundle savedInstanceState) {
        super.onCreate(savedInstanceState);
        setContentView(R.layout.act_main);
        initWidget();
    }
    private void initWidget()
    {
        Button btn = (Button)findViewById(R.id.btn_click);      //获取按钮
        btn.setOnClickListener(new OnClickListener()
        {
            public void onClick(View v) {
                Intent intent = new Intent();      //定义一个Intent对象
                //没有具体指定需要启动的Activity，只是设置了action
                intent.setAction("com.demo.action.show");
                startActivity(intent);
            }
        });
    }
}
```

ShowActivity 代码：

```
public class ShowActivity extends Activity {
```

```
protected void onCreate(Bundle savedInstanceState) {
    super.onCreate(savedInstanceState);
    setContentView(R.layout.act_show);
    Intent intent = getIntent();            //获取Intent的对象
    String action = intent.getAction();     //获取action内容
    TextView txtView = (TextView)findViewById(R.id.txt_show);
    txtView.setText("action is: " + action);    //显示action内容
}
```

ShowActivity 的配置：

```
<activity android:name="com.demo.intent.ShowActivity">
        <intent-filter>
            <action android:name="com.demo.action.show"/>
            <category android:name="android.intent.category.DEFAULT"/>
        </intent-filter>
</activity>
```

- 案例分析

MainActivity 的内容比较简单，仅仅定义了一个按钮对象，单击之后启动另外一个 Activity。从 MainActivity 代码中可以看出，在单击按钮之后，没有为 Intent 对象指定具体需要启动的组件名，仅仅设置了 action 属性，即 MainActivity 并不关心将要启动谁。ShowActivity 中定义了一个 TextView 用于显示获取的 action 名。另外 ShowActivity 的配置增加了一个 action 属性。

Android 系统也提供了一系列标准的 Action，常见的 Action 如表 4.1 所示。

表 4.1　系统标准 Action 举例

常量定义	对应字符串	说明
Intent.ACTION_CALL	android.intent.action.CALL	向指定用户拨打电话
Intent.ACTION_EDIT	android.intent.action.EDIT	编辑指定的数据
Intent.ACTION_MAIN	android.intent.action.MAIN	程序的入口，即程序的启动页面
Intent.ACTION_VIEW	android.intent.action.VIEW	查看指定的数据
Intent.ACTION_SEND	android.intent.action.SEND	发送数据
Intent.ACTION_DIAL	android.intent.action.DIAL	显示设备的拨号页面

2. data 属性

data 属性通常与 action 属性一起使用，用于向 action 属性提供操作所需要的数据。data 由两部分组成：MIME type 和 URI。一个 URI 对象通常以如下的字符串形式表示：scheme://host:port/path。MIME type 用于明确指定数据的类型。在这里需要注意 Intent.setType()、Intent.setData()和 Intent.setData

AndType()这 3 个方法，前两个方法会互相覆盖，调用 setType()方法之后，系统会先设置 mimeType，然后将 data 置为 null。同样地，如果调用 SetData()方法，系统会先设置 URI，然后将 mimeType 置为 null。如果想同时设置 URI 和 mimeType，可以调用 setDataAndType()。

3. category 属性

大多数 Intent 不需要添加 category 属性，通过 action 和 data 属性已经可以准确地表达一个完整的意图了。但有些时候，为了使匹配更加明确，可以添加 category 属性作为附加信息。对于一个 Intent，只能指定一个 action 属性，但是可以指定多个 category 属性。表 4.2 列出了 Android 系统自带的一些标准 category 属性。

表 4.2　系统标准 category 属性举例

常量定义	对应字符串	说明
CATEGORY_BROWSABLE	android.intent.category.BROWSABLE	该 Activity 可以安全地被浏览器调用
CATEGORY_HOME	android.intent.category.HOME	该 Activity 随系统启动
CATEGORY_LAUNCHER	android.intent.category.LAUNCHER	应用启动时第一个启动的 Activity
CATEGORY_PREFERENCE	android.intent.category.PREFERENCE	该 Activity 是参数面板

4.2　BroadCastRecevier

BroadCast Receiver

BroadCastRecevier 也是 Android 四大组件之一，用于在系统范围内接收广播通知，例如电池电量不足、数据下载完成等。在 BroadCastRecevicer 中一般会做一些轻量级的处理。

4.2.1　广播机制介绍

Android 这种被动等待接收消息的机制类似于广播的处理，广播发送者发送一条消息出去，广播接收者接收到消息之后做对应的处理，BroadCastRecevicer 就是广播接收者。了解设计模式的读者也可以将这种处理方式和观察者模式做比较。具体的实现流程如下。

（1）BroadCastReceiver 向系统进行注册。
（2）广播发送者发送广播。
（3）系统查找符合相应条件的 BroadCastReceiver，将广播内容发送到 BroadCastReceiver 相应的消息循环队列中。
（4）回调 BroadCastReceiver 中的 onReceive()方法。

广播发送者和接收者之间的消息传递通过 Intent 实现。

4.2.2　静态注册

BroadCastReceiver 在使用之前需要向系统注册，有两种注册方式：静态注册和动态注册。静态注册是在 AndroidManifest.xml 中进行配置，BroadCastRecevier 的配置与 Activity 类似。下面我们通过一个实例说明一下 BroadCastReceiver 的使用方法。

案例 4.2　BroadCastReceiver 的使用

在页面中显示一个按钮，单击之后，调用 sendBroadcast()方法发送广播，参数为 Intent。运行结果如图 4.2 所示，单击按钮之后，控制台打印出了接收到的内容。

图 4.2　静态注册广播的运行结果

- 案例代码

MyReceiver 代码（自定义广播接收者）：

```
//定义一个广播接收者继承自BroadCastReceiver，用于接收广播消息
public class MyReceiver extends BroadcastReceiver
{
    public void onReceive(Context context, Intent intent)    //实现onReceive回调方法
    {
        if(null == intent)
        {
            return;
        }
        String action = intent.getAction();       //取出Intent的具体action
        //判断action和配置的是否一致
        if(!TextUtils.isEmpty(action) && "android.intent.action.receiverdemo".equals(action))
        {
            String msg = intent.getStringExtra("msg");    //取出Intent携带的数据
            Log.i("MyReceiver", "msg:" + msg);    //打印接收到的内容
        }
    }
}
```

首先自定义一个广播接收者，实现其中的 onReceive()方法，在其中做相应的处理。onReceive()回调方法有两个参数，一个是发送广播上下文的 Context，一个是携带广播数据的 Intent。

在 AndroidManifest.xml 中注册广播：

```xml
<receiver android:name="com.demo.receiver.MyReceiver">
    <intent-filter>
        <action android:name="android.intent.action.receiverdemo"/>
    </intent-filter>
</receiver>
```

可以看到，广播接收者的静态注册和 Activity 的配置类似，需要定义一个 action 用于接收特定的广播。

MainActivity 代码：

```java
public class MainActivity extends Activity
{
    protected void onCreate(Bundle savedInstanceState)
    {
        super.onCreate(savedInstanceState);
        setContentView(R.layout.activity_main);
        initWidget();
    }
    private void initWidget()
    {
        Button btnSend = (Button)findViewById(R.id.btn_send);
        btnSend.setOnClickListener(new OnClickListener()
        {
            public void onClick(View v)
            {
                Intent intent = new Intent();    //定义Intent
//设置Intent的action属性
                intent.setAction("android.intent.action.receiverdemo");
                intent.putExtra("msg", "this is a broadcast");    //携带数据
                MainActivity.this.sendBroadcast(intent);    //发送广播
            }
        });
    }
}
```

4.2.3 动态注册

除了在 AndroidManifest.xml 中静态配置广播之外，还可以在代码中动态地注册广播。下面我们通过一个案例来了解一下。

案例 4.3　动态注册广播

本例通过代码说明动态注册广播的方式。

- **案例代码**

MainActivity 代码：

```java
public class MainActivity extends Activity
{
    private MyReceiver mReceiver = null;
    protected void onCreate(Bundle savedInstanceState)
    {
        super.onCreate(savedInstanceState);
        setContentView(R.layout.activity_main);
        initWidget();
        mReceiver = new MyReceiver();      //定义一个广播接收者的对象
        //定义一个IntentFilter，指定该广播接收者可以匹配的action
        IntentFilter filter = new IntentFilter("android.intent.action.receiverdemo");
        registerReceiver(mReceiver, filter);     //注册广播
    }
    protected void onDestroy()
    {
        if(null != mReceiver)
        {
            unregisterReceiver(mReceiver);      //反注册广播接收者
            mReceiver = null;
        }
    }
    private void initWidget()
    {
        Button btnSend = (Button)findViewById(R.id.btn_send);
        btnSend.setOnClickListener(new OnClickListener()
        {
            public void onClick(View v)
            {
                Intent intent = new Intent();     //定义Intent
                intent.setAction("android.intent.action.receiverdemo");     //设置Intent的action属性
                intent.putExtra("msg", "this is a broadcast");    //携带数据
                MainActivity.this.sendBroadcast(intent);      //发送广播
```

```
            }
        });
    }
}
```

- **案例分析**

从代码中可以看到，在 Activity 的 onCreate()方法中定义了一个 MyReceiver 的对象，然后定义一个 IntentFilter 用于指明广播接收者可以匹配的 action，然后通过 registerReceiver()方法注册广播。需要注意的是，在退出界面时，需要调用 unregisterReceiver()方法反注册广播接收者。

> 对于开发者来说，一个类中的方法最好成对出现，例如有一个灯泡管理类，如果对外提供了开灯方法，最好同时对外提供一个关灯方法。对于 Activity，生命周期的方法也都是成对出现，onCreate()方法是 Activity 生命周期开始的方法，可以在其中做一些初始化的工作，广播的注册过程就可以放在其中。onDestroy()方法是 Activity 生命周期结束的方法，可以在其中做一些资源释放和回收，广播的反注册过程就可以放在其中。

4.2.4 系统广播介绍

除了自定义的广播之外，BroadCastReceiver 还有一个很重要的作用是接收系统广播。系统在执行一些特定的操作或者在一些特定的状态下会发送系统广播。例如，系统在刚开机的时候会发送开机广播，在电池电量低的时候也会发送广播。表 4.3 列出了一些常见的系统广播 Action 常量。

表 4.3 系统广播 Action 举例

常量定义	说明
ACTION_TIME_CHANGED	系统时间发生改变
ACTION_TIMEZONE_CHANGED	系统时区发生改变
ACTION_BOOT_COMPLETED	系统开机完成
ACTION_PACKAGE_ADDED	系统添加包，一般在安装新应用之后
ACTION_PACKAGE_REMOVED	系统删除包，一般在应用卸载之后
ACTION_BATTERY_CHANGED	电池电量发生改变
ACTION_BATTERY_LOW	电池电量低
ACTION_POWER_CONNECTED	设备连接上电源
ACTION_POWER_DISCONNECTED	设备与电源断开
ACTION_SHUTDOWN	设备关机

下面我们通过一个实例说明一下系统广播的使用。

案例 4.4 通过接收系统广播提示用户充电

在使用手机的过程中，如果手机的电池电量过低，一般都会提示用户需要充电，我们可以通过接收系统广播实现这个功能。

- **案例代码**

MyReceiver 代码：

```java
//定义一个广播接收者继承自BroadCastReceiver，用于接收广播消息

public class MyReceiver extends BroadcastReceiver
{
    public void onReceive(Context context, Intent intent)     //实现onReceive回调方法
    {
        Bundle bundle = intent.getExtras();         //获取Intent携带的数据
        int curBarray = bundle.getInt("level");     //获取当前电量
        int totalBarray = bundle.getInt("scale");   //获取总电量
        if(curBarray * 1.0 / totalBarray < 0.20)    //电量小于20%，提示用户充电
        {
            Toast.makeText(context, "电池电量低，请充电！", Toast.LENGTH_LONG).show();
        }
    }
}
```

广播的注册：

```xml
<receiver android:name="com.demo.receiver.MyReceiver">
    <intent-filter>
        <action android:name="android.intent.action.BATTERY_CHANGED"/>
    </intent-filter>
</receiver>
```

从上述实例可以看出，系统广播的使用与自定义的广播类似，需要定义一个广播接收者处理接收到广播之后的操作，然后对广播进行注册，注意注册使用的 action 与表 4.3 中列出的系统广播的 action 要一致。

4.3 本章小结

本章我们首先介绍了 Intent 和 IntentFilter 的概念，Intent 一般用于指示想要启动的组件和传递参数，第 3 章关于 Activity 的小练习也显示了 Intent 的具体使用方式。然后介绍了 Android 的广播机制，主要包括广播接收者的创建和注册。最后介绍了系统在发生一些行为时会发送的系统广播，用户可以自定义广播接收者接收这些信息并做相应的处理。广播接受者一般会和其他组件结合使用，后续我们将在第 7 章通过一个小练习学习一下如何接收系统广播。

第5章

数据存储

■ 所有的应用程序都会产生数据的存储和读写。例如，手机游戏可以保存用户当前玩到的关卡和分数，音乐播放器或视频播放软件会记录用户的使用习惯，这些配置在用户关机重启之后仍然会保存，说明数据都保存在了文件中。Android 为开发者封装了大量的 API 可以进行数据操作，对轻量级数据的读写可以使用 SharedPrepfereces。如果想访问手机存储器上的数据，可以使用 IO 流操作。另外，应用程序如果需要操作大量的数据，可以使用 Android 内置的 SQLite 数据库。

5.1 SharedPreferences

SharedPreferences 和 File 存储

SharedPreferences 用于保存应用程序中少量的数据,例如音乐播放器的播放模式是顺序播放还是随机播放,手机游戏是否打开音效振动等,这些数据都可以使用 SharedPreferences 保存。

5.1.1 SharedPreferences 与 Editor 简介

SharedPreferences 提供了一个通用的框架,以便开发者可以以键值对的形式将数据持久化,值得注意的是,SharedPreferences 只能保存原始的数据类型:布尔值、浮点值、整型值、长整型和字符串。要获取 SharedPreferences 对象,可以调用 Context 提供的 getSharedPreferences(String name, int mode) 方法,该方法的第 1 个参数指定了创建的 SharedPreferences 文件名,第 2 个参数可以取如下几个值。

(1) MODE_PRIVATE:SharedPreferences 的默认值,指定创建的文件只能本应用程序访问。

(2) MODE_WORLD_READABLE:指定创建的文件可以被其他应用程序读。

(3) MODE_WORLD_WRITEABLE:指定创建的文件可以被其他应用程序读写。

SharedPreferences 接口本身并没有写数据能力,而是通过 Editor 对象实现,可以通过调用 SharedPreferences 的 edit()方法获取 Editor 对象,Editor 对外提供了大量写数据的方法,具体如表 5.1 所示。

表 5.1 Editor 提供的方法

方法名	方法说明
clear()	清除 SharedPreferences 文件中所有的数据
commit()	Editor 完成编辑后,调用该方法提交修改
putBoolean(String key, boolean value)	保存一个布尔型变量
putFloat(String key, float value)	保存一个浮点型变量
putInt(String key, int value)	保存一个整型变量
putLong(String key, long value)	保存一个长整型变量
putString(String key, String value)	保存一个字符串变量
remove(String key)	从 SharedPreferences 文件中移除键 key 对应的值

5.1.2 SharedPreferences 存储的位置和格式

下面我们通过一个实例说明一下 SharedPreferences 的使用方法。

案例 5.1 使用 SharedPreferences 存储数据

本例在界面上显示两个按钮,单击其中一个按钮,向 SharedPreferences 中写入一个随机值。单击另外一个按钮,在页面上显示刚才写入的数值,运行结果如图 5.1 所示。

第 5 章
数据存储

图 5.1 SharedPreferences 读写数据

- **案例代码**

MainActivity 代码：

```java
public class MainActivity extends Activity {
    private SharedPreferences mPreferences;        //定义一个SharedPreferences对象
    @Override
    protected void onCreate(Bundle savedInstanceState) {
        super.onCreate(savedInstanceState);
        setContentView(R.layout.activity_main);
        init();
    }
    private void init()
    {
        //调用getSharedPreferences方法获取Sharepreferences的具体实例
        mPreferences = getSharedPreferences("demoshareprefeces", MODE_PRIVATE);
        Button btnWrite = (Button)findViewById(R.id.btn_write);      //获取写按钮
        Button btnRead = (Button)findViewById(R.id.btn_read);        //获取读按钮
        final TextView txtInfo = (TextView)findViewById(R.id.txt_info);
        btnWrite.setOnClickListener(new View.OnClickListener() {
            public void onClick(View v) {
                SharedPreferences.Editor editor = mPreferences.edit();     //获取Editor对象
                int value = new Random().nextInt(1000);       //生成一个随机数
                editor.putInt("random", value);      //将随机数写入SharedPreferences中
                editor.commit();     //提交Editor所做的编辑
```

```
            }
        });
        btnRead.setOnClickListener(new View.OnClickListener() {
            @Override
            public void onClick(View v) {
                int value = mPreferences.getInt("random", 0);      //调用getXXX方法读取数值
                txtInfo.setText("The number is: " + value);        //显示读取到的数值
            }
        });
    }
}
```

- **案例分析**

本例首先调用了getSharedPreferences()方法获取了一个SharedPreferences对象，然后调用editor()方法获取 Editor 对象，将生成的随机数保存到 SharedPreferences 中。在读取的时候，调用 getXXX()方法获取指定 key 对应的数值。

getSharedPreferences()方法不仅会返回一个SharePreferences对象，还会创建一个文件用于保存数据，该文件存储在设备的"/data/data/应用包名/shared_prefs"路径下，打开AndroidStudio的DDMS窗口可以看到，如图 5.2 所示，demosharepfeces.xml 即为创建的 SharedPreferences 文件。

图 5.2　SharedPrefences 文件保存的位置

将该文件 pull 到电脑上，使用记事本打开，可以看到其中的内容如图 5.3 所示，以键值对的形式保存。

```
1 <?xml version='1.0' encoding='utf-8' standalone='yes' ?>
2 <map>
3     <int name="random" value="765" />
4 </map>
```

图 5.3　SharedPrefences 文件内容

5.2　File 存储

Android 中的文件存储支持使用 Java 原生的 File 类和一些文件流进行访问，但是 Android 的存储

规则和其他系统有所不同，分为内部存储和外部存储。内部存储位于系统中一个特定的位置，写入其中的文件不能在应用之间共享，当应用被卸载后，处在内部存储中的对应文件也被删除。外部存储的路径可以通过 Android 提供的接口获取到。本节我们将介绍如何通过 File 类和相关的操作类读写 Android 的内部和外部存储文件。

5.2.1 读写内部存储

在开发中，可以将文件直接保存在设备的内部存储空间中。默认地，保存在内部存储空间中的文件是私有的，其他程序没有权限访问该文件。当本应用被卸载时，这些内部存储的文件也会被移除。内部存储的文件通过 IO 流实现读写。Android 提供了以下两个方法获取输入流和输出流。

（1）FileInputStream openFileInput(String name)：获取内部存储中 name 文件对应的输入流。

（2）FileOutputStream openFileOutput(String name, int mode)：获取内部存储中 name 文件对应的输出流，mode 指定了打开文件的模式，可以取如下值。

- MODE_PRIVATE：该文件只能被当前程序读写。
- MODE_APPEND：以追加的方式打开文件。

下面我们通过一个案例说明一下读写内部存储的方法。

案例 5.2　使用内部存储

本例功能与案例 5.1 类似，只不过将数据的读写由 SharedPreferences 改为读写内部存储。运行结果如图 5.4 所示。内部存储的文件存储在 "/data/data/应用包名/files" 路径下，如图 5.5 所示。

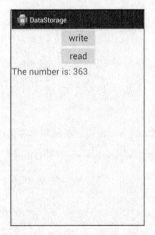

图 5.4　读写内部存储文件运行结果

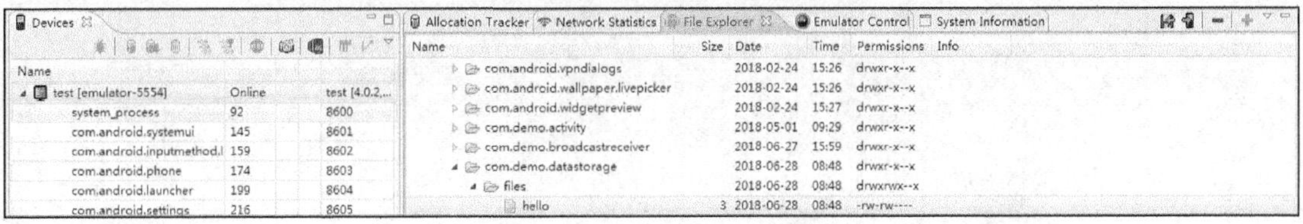

图 5.5　内部存储保存文件的位置

- **案例代码**

MainActivity 代码：

```java
public class MainActivity extends Activity {
    private final String FILE_NAME = "hello";     //将文件名定义为常量
    protected void onCreate(Bundle savedInstanceState) {
        super.onCreate(savedInstanceState);
        setContentView(R.layout.activity_main);
        init();
    }
    private void init()
    {
        Button btnWrite = (Button)findViewById(R.id.btn_write);
        Button btnRead = (Button)findViewById(R.id.btn_read);
        final TextView txtInfo = (TextView)findViewById(R.id.txt_info);
        btnWrite.setOnClickListener(new View.OnClickListener() {
            @Override
            public void onClick(View v) {
                int value = new Random().nextInt(1000);     //生成一个随机值
                write(value);     //调用方法将随机值写入内部存储中
            }
        });
        btnRead.setOnClickListener(new View.OnClickListener() {
            @Override
            public void onClick(View v) {
                int value = read();     //调用方法从内部存储中读取随机值
                txtInfo.setText("The number is: " + value);     //显示
            }
        });
    }
    private void write(int value)
    {
        try
        {
            //获取输出流
            FileOutputStream fos = openFileOutput(FILE_NAME, MODE_PRIVATE);
```

```
                fos.write(String.valueOf(value).getBytes());    //将数值写入文件中
                fos.close();        //关闭流
            } catch (Exception e) {
                e.printStackTrace();
            }
        }
        private int read()
        {
            try {
                FileInputStream fis = openFileInput(FILE_NAME);      //获取输入流
                DataInputStream dis = new DataInputStream(fis);
                int value = dis.readInt();      //读取数值
                dis.close();        //关闭流
                fis.close();
                return value;
            } catch (Exception e) {
                e.printStackTrace();
                return 0;
            }
        }
    }
```

5.2.2 读写外部存储

一般情况下，手机的内部存储空间有限，如果需要读写大文件数据，Android 提供了相关访问外部存储的方法。读写外部存储需要在 AndroidManifest.xml 文件中配置权限，如下所示：

```
<manifest ...>
    <uses-permission android:name="android.permission.WRITE_EXTERNAL_STORAGE" />
    ...
</manifest>
```

读写外部存储步骤一般如下。

（1）调用 Environment 的 getExternalStorageState()方法判读外部存储设备是否可用，如果外部存储设备可读写，则 Environment.getExternalStorageState().equals(Environment.MEDIA_MOUNTED)返回 true。

（2）调用 Environment 的 getExternalStorageDirectory()方法获取外部存储器的目录。

（3）使用 FileInputStream、FileOutpuStream、FileWriter、FileReader 等流操作文件读写外部存储上

的文件。

5.3 SQLite 数据库

SQLite 数据库

在应用程序的开发过程中，如果需要存取的数据量比较大，可以考虑使用数据库。Android 系统集成了一个轻量级的数据库——SQLite。它不像 Oracle、SQL Server、MySQL 那样需要单独安装，SQLite 只是一个文件，操作比较简单，Android 提供了大量的 API 支持 SQLite 读写数据。

5.3.1 SQLiteDatabase 简介

SQLiteDatabase 代表一个数据库，应用程序获取了指定数据库的 SQLiteDatabase 对象之后，可以调用一系列的方法操作数据库。表 5.2 列出了 SQLiteDatabase 类中一些常用的方法。

表 5.2 SQLiteDatabase 类提供的方法

方法和对应的说明
openDatabase(String path, SQLiteDatabase.CursorFactory factory, int flags) 打开 path 文件代表的 SQLite 数据库
openOrCreateDatabase(String path, SQLiteDatabase.CursorFactory factory) 打开 path 文件代表的 SQLite 数据库，如果不存在，则创建
execSQL(String sql) 执行 sql 语句
execSQL(String sql, Object[] bindArgs) 执行带占位符的 sql 语句
insert(String table, String nullColumnHack, ContentValues values) 向表中插入一条数据
delete(String table, String whereClause, String[] whereArgs) 删除表中指定的数据
update(String table, ContentValues values, String whereClause, String[] whereArgs) 更新表中指定的数据
query(String table, String[] columns, String selection, String[] selectionArgs, String groupBy, String having, String orderBy) 按一定条件查询表中的数据
query(String table, String[] columns, String selection, String[] selectionArgs, String groupBy, String having, String orderBy, String limit) 按一定条件查询表中的数据，并且控制查询的个数。多用于分页
rawQuery(String sql, String[] selectionArgs) 执行带占位符的查询语句
beginTransaction() 开始一个事务

续表

方法和对应的说明
setTransactionSuccessful() 设置当前的事务为成功
endTransaction() 结束一个事务

5.3.2 创建数据库和表

在上一节我们介绍了 SQLiteDatabase 的作用，并提供了相关的方法获取 SQLiteDatabase 对象，但是在实际使用过程中，通常要通过 SQLiteOpenHelper 类获取 SQLiteDatabase 对象，下面我们通过一个具体实例看一下数据库和表的具体创建。

案例 5.3　创建表，存储学生考试成绩

本例创建一个学生表，用于存储学生的考试成绩，字段包括学号、姓名、性别、考试成绩。

- 案例代码

DBHelper 代码：

```java
public class DBHelper extends SQLiteOpenHelper {
    private static final int DB_VERSION = 1;        //数据库版本号
    private static final String DB_NAME = "student.db";     //数据库名称
    private static final String TABLE_NAME = "score";      //创建的表的名称
    public DBHelper(Context context)     //接收Context参数的构造方法
    {
        super(context, DB_NAME, null, DB_VERSION);       //调用父类构造方法创建数据库
    }
    public void onCreate(SQLiteDatabase db) {   //数据库第一次被创建时会回调onCreate()
        String sql = "create table if not exists " + TABLE_NAME + " (stuNo text primary key, stuName text, stuSex text, stuScore integer)";      //创建表的语句
        db.execSQL(sql);    //执行创建表的sql语句
    }
    @Override
    public void onUpgrade(SQLiteDatabase db, int oldVersion, int newVersion) {
    }
}
```

- 案例分析

本例定义了一个 DBHelper 类继承自 SQLiteOpenHelper，调用父类的构造方法 super(context, DB_

NAME, null, DB_VERSION)会自动创建名为 DB_NAME 的数据库,数据库第一次被创建时会毁掉 onCreate()方法,可以在其中执行创建表的操作,创建的数据库存储在"/data/data/应用包名/databases"路径下,如图 5.6 所示。

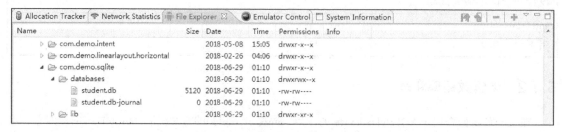

图 5.6　sqlite 数据库保存的位置

5.3.3　操作 SQLite 常用类

数据的读写一般就是增、删、改、查 4 个操作,下面我们通过一个实例看一下如何对 SQLite 数据库进行访问。

案例 5.4　访问 SQLite 数据库,修改学生成绩表

本例在界面上定义 4 个按钮,分别触发增加数据、删除数据、修改数据、查询数据操作,运行结果如图 5.7 所示。

图 5.7　触发访问数据库的 4 个按钮

- 案例代码

MainActivity 代码:

```
public class MainActivity extends Activity implements OnClickListener {
```

```java
private DBMgr mDBMgr = null;    //用于封装对SQLite数据库的访问
protected void onCreate(Bundle savedInstanceState) {
    super.onCreate(savedInstanceState);
    setContentView(R.layout.activity_main);
    init();
}
private void init()
{
    mDBMgr = new DBMgr(this);   //初始化
    Button btnAdd = (Button)findViewById(R.id.btn_add);   //定义4个按钮
    Button btnDelete = (Button)findViewById(R.id.btn_delete);
    Button btnUpdate = (Button)findViewById(R.id.btn_update);
    Button btnQuery = (Button)findViewById(R.id.btn_query);
    btnAdd.setOnClickListener(this);
    btnDelete.setOnClickListener(this);
    btnUpdate.setOnClickListener(this);
    btnQuery.setOnClickListener(this);
}
public void onClick(View v) {
    switch(v.getId())
    {
    case R.id.btn_add:   //增加数据
    {
        Student stu1 = new Student("201701", "aaa", "男", 95);
        Student stu2 = new Student("201702", "aaa", "女", 99);
        Student stu3 = new Student("201703", "aaa", "男", 97);
        List<Student> list = new ArrayList<Student>();
        list.add(stu1);
        list.add(stu2);
        list.add(stu3);
        mDBMgr.add(list);
        break;
    }
    case R.id.btn_delete:   //删除数据
    {
```

```
            mDBMgr.delete("201703");
            break;
        }
        case R.id.btn_update:      //更新数据
        {
            mDBMgr.update("201701", 90);
            break;
        }
        case R.id.btn_query:      //查询数据
        {
            List<Student> list = mDBMgr.query();
            for(Student stu : list)
            {
                Log.i("Student", stu.toString());
            }
            break;
        }
        default:
            break;
        }
    }
    protected void onDestroy()
    {
        mDBMgr.finish();       //资源回收工作
        super.onDestroy();
    }
}
```

具体访问数据库的操作封装在了类 DBMgr 中，DBMgr 的具体代码如下。

DBMgr 代码：

```
public class DBMgr {
    private SQLiteDatabase mDB = null;    //定义一个SQLiteDatabase的对象
    /**
     * 构造函数，做一些初始化工作
     * @param context
     */
```

```java
public DBMgr(Context context)
{
    DBHelper helper = new DBHelper(context);
    mDB = helper.getWritableDatabase();        //获取一个SQLiteDatabase类的实例
}
/**
 * 关闭数据库操作
 */
public void finish()
{
    if(null != mDB)
    {
        mDB.close();
        mDB = null;
    }
}
/**
 * 向数据库中添加数据
 * @param students
 */
public void add(List<Student> students)
{
    if(null == students || students.isEmpty())
    {
        return;
    }
    for(Student stu : students)
    {
        ContentValues values = new ContentValues();
        values.put("stuNo", stu.getStuNo());
        values.put("stuName", stu.getStuName());
        values.put("stuSex", stu.getStuSex());
        values.put("stuScore", stu.getStuScore());
        mDB.insert("score", null, values);        //执行添加操作
    }
}
/**
```

```java
 * 删除指定学号的学生信息
 * @param stuNo 学生学号
 */
public void delete(String stuNo)
{
    mDB.delete("score", "stuNo=?", new String[]{stuNo});
}
/**
 * 更新指定学号的学生分数
 * @param stuNo 学生学号
 * @param score 分数
 */
public void update(String stuNo, int score)
{
    String sql = "update score set stuScore=" + score + " where stuNo=" + stuNo;
    mDB.execSQL(sql);
}
/**
 * 查询所有学生信息
 * @return 返回查询结果
 */
public List<Student> query()
{
    List<Student> list = new ArrayList<Student>();
    //执行查询语句
    Cursor cursor = mDB.query("score", null, null, null, null, null, null);
    while(cursor.moveToNext())      //遍历查询结果
    {
        Student stu = new Student();
        stu.setStuNo(cursor.getString(cursor.getColumnIndex("stuNo")));
        stu.setStuName(cursor.getString(cursor.getColumnIndex("stuName")));
        stu.setStuSex(cursor.getString(cursor.getColumnIndex("stuSex")));
        stu.setStuScore(cursor.getInt(cursor.getColumnIndex("stuScore")));
        list.add(stu);
    }
    return list;
}
```

}

- **案例分析**

在 DBMgr 的构造函数中,调用了 SQLiteOpenHelper 类的 getWritableDatabase()方法获取了一个 SQLiteDatabase 对象,用于操作数据库。

1. 查询操作

调用了表 5.2 中所示的 query(String table, String[] columns, String selection, String[] selectionArgs, String groupBy, String having, String orderBy)的方法,这里需要查询所有数据,所以除了第一个参数传表的名称之外,其他参数均传 null。查询结果是一个 Cursor 对象,Cursor 类似于一个集合,可以遍历取出其中的数据。

2. 添加操作

调用了表 5.2 中所示的 insert(String table, String nullColumnHack, ContentValues values)的方法,数据是以键值对的形式保存在 ContentValues 对象中,然后插入,插入 3 条数据之后的查询结果如图 5.8 所示。

```
I  06-29 01:33:43.081    8332    8332    com.demo.sqlite    Student    stuNo: 201801; stuName: aaa; stuSex:男; stuScore: 95
I  06-29 01:33:43.081    8332    8332    com.demo.sqlite    Student    stuNo: 201802; stuName: aaa; stuSex:女; stuScore: 99
I  06-29 01:33:43.081    8332    8332    com.demo.sqlite    Student    stuNo: 201803; stuName: aaa; stuSex:男; stuScore: 97
```

图 5.8　SQLite 插入数据结果

3. 修改操作

调用了表 5.2 中所示的 update(String table, ContentValues values, String whereClause, String[] whereArgs)方法,按指定条件更新数据,更新后的结果如图 5.9 所示,将学号为 201801 的学生的成绩改为了 90 分。

```
I  06-29 01:34:38.460    8332    8332    com.demo.sqlite    Student    stuNo: 201801; stuName: aaa; stuSex:男; stuScore: 90
I  06-29 01:34:38.460    8332    8332    com.demo.sqlite    Student    stuNo: 201802; stuName: aaa; stuSex:女; stuScore: 99
I  06-29 01:34:38.460    8332    8332    com.demo.sqlite    Student    stuNo: 201803; stuName: aaa; stuSex:男; stuScore: 97
```

图 5.9　SQLite 更新数据结果

4. 删除操作

调用了表 5.2 中的 delete(String table, String whereClause, String[] whereArgs)方法,将学号为 201803 的学生的数据删除,运行结果如图 5.10 所示。

```
I  06-29 01:35:12.521    8332    8332    com.demo.sqlite    Student    stuNo: 201801; stuName: aaa; stuSex:男; stuScore: 90
I  06-29 01:35:12.521    8332    8332    com.demo.sqlite    Student    stuNo: 201802; stuName: aaa; stuSex:女; stuScore: 99
```

图 5.10　SQLite 删除数据结果

5.3.4　事务

在数据库中,事务是指一系列操作的集合,这些操作要么全部成功执行,要么全部不执行。一个典型的例子是银行的转账工作，它包含两个操作：（1）从一个账户扣款；（2）另一个账户增加相应的金额。很明显，这两个操作必须全部成功执行或者都不执行。SQLite 数据库对事务的概念也做了相应的支持，在操作的开始处调用 beginTransaction()开启一个事务，当一系列的操作完成后，调用 setTransactionSuccessful()函数设置事务成功，最后调用 endTransaction()结束一个事务。案例 5.4 中添加 3 条学生信息的操作可以使用事务的概念，代码如下：

```
/**
 * 向数据库中添加数据
 * @param students
 */
public void add(List<Student> students)
{
    if(null == students || students.isEmpty())
    {
        return;
    }
    mDB.beginTransaction();    //开始一个事务
    for(Student stu : students)
    {
        ContentValues values = new ContentValues();
        values.put("stuNo", stu.getStuNo());
        values.put("stuName", stu.getStuName());
        values.put("stuSex", stu.getStuSex());
        values.put("stuScore", stu.getStuScore());
        mDB.insert("score", null, values);    //执行添加操作
    }
    mDB.setTransactionSuccessful();    //操作完成之后将事务设置为成功
    mDB.endTransaction();    //结束一个事务
}
```

5.4 本章小结

本章我们主要介绍了 Android 中的数据存储，首先介绍了 Android 提供的一个通用框架 SharedPreferences，该框架使得开发者能够以键值对的形式保存一些数据。然后介绍了读取 Android 系统内部存储和外部存储的方式，可以调用系统提供的接口获取文件的输入输出流。当数据量比较大而且是结构化的数据时，数据库可能是一种比较好的选择，Android 集成了一个轻量级的 SQLite 数据库，在本章我们介绍了 SQLite 数据库的一些基础知识和 Android 如何操作 SQLite 数据。

5.5 小练习

下面我们通过一个小练习来具体熟悉一下对于 Android 系统中的文件访问，本练习实现了一个文件浏览器。刚进入应用的时候，会以列表的形式排序展示系统中存储的文件，排序规则为：文件夹在前面，文件在后面，如果同为文件夹或文件，则按字母顺序排列。文件夹和文件显示不同的图标，单击文件夹，可以进入下级目录并展示。应用顶部显示当前所在目录，如图 5.11 和图 5.12 所示。

图 5.11 文件列表

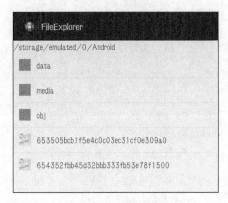

图 5.12 下级目录

主页首先获取系统外部存储的根目录，然后通过 listview 控件列表展示。当单击某个文件夹时，获取单击文件夹的文件路径，通过 listFiles()方法获取该路径下的子文件，并根据自定义的比较器对结果进行排序，然后更新 listview 的数据，刷新显示，具体代码如下。

MainActivity 代码：

```
public class MainActivity extends Activity {
    private ListViewAdapter mAdpter;
    private TextView mTitle;
    private FileMgr fileMgr;


    @Override
    protected void onCreate(Bundle savedInstanceState) {
        super.onCreate(savedInstanceState);
        setContentView(R.layout.activity_main);
```

```java
        //用于显示当前文件的路径
        mTitle = (TextView)findViewById(R.id.txt_view);
        //获取listview控件
        ListView listView = (ListView)findViewById(R.id.list_view);
        //自定义的Adapter
        mAdpter = new ListViewAdapter(this);
        init();
        listView.setAdapter(mAdpter);
        listView.setOnItemClickListener(new OnItemClickListener(){
            public void onItemClick(AdapterView<?> parent, View view, int pos,
                    long id) {
                //获取当前单击的文件
                File file = mAdpter.getItem(pos);
                //更新显示下级目录
                change(file);
            }
        });
    }

    private void init()
    {
        fileMgr = new FileMgr();
        //判断外部存储是否可用
        if(!Environment.getExternalStorageState().equals(Environment.MEDIA_MOUNTED))
        {
            Toast.makeText(this, "the external storage is not avalible", Toast.LENGTH_SHORT).show();
            return;
        }
        File rootFile = Environment.getExternalStorageDirectory();
        mTitle.setText(rootFile.getAbsolutePath());
        //获取当前目录的子文件列表
        List<File> files = fileMgr.getSubFiles(rootFile);
        mAdpter.updateFiles(files);
    }
    private void change(File file)
    {
```

```
            if(!file.isDirectory())     //如果不是文件夹，直接返回
            {
                return;
            }
            //更新路径的显示
            mTitle.setText(file.getAbsolutePath());
            //获取新的文件列表
            List<File> files = fileMgr.getSubFiles(file);
            //更新文件和视图显示
            mAdpter.updateFiles(files);
            mAdpter.notifyDataSetChanged();
        }
    }
```

在 MainActivity 中，首先根据 Environment.getExternalStorageState()方法判断外部存储是否可用，如果可用，获取外部存储的根目录，然后调用文件管理类 FileMgr 的方法获取子文件列表，将文件列表设置到 listView 自定义的 Adapter 中。在 onCreate()方法中，对 listview 的 item 设置了单击的监听事件，当 item 被单击后，获取对应的文件对象。如果该文件属于文件夹，则获取子目录，并调用 Adapter 的 notifyDataSetChanged()方法刷新页面显示。其中自定义 Adapter 和文件管理类的代码如下。

ListViewAdapter 代码：

```
public class ListViewAdapter extends BaseAdapter{

    private LayoutInflater flater;

    private List<File> mDatas;

    public ListViewAdapter(Context context)
    {
        flater = LayoutInflater.from(context);
        mDatas = new ArrayList<File>();
    }

    public void updateFiles(List<File> files)
    {
        mDatas.clear();
        mDatas.addAll(files);
    }
```

```java
public int getCount() {
    return (null == mDatas || mDatas.isEmpty()) ? 0 : mDatas.size();
}

@Override
public File getItem(int pos) {
    if(null != mDatas && pos < mDatas.size())
    {
        return mDatas.get(pos);
    }
    return null;
}

@Override
public long getItemId(int position) {
    return 0;
}

@Override
public View getView(int position, View convertView, ViewGroup parent) {
    if(null == convertView)
    {
        //为每一个子项加载布局
        convertView = flater.inflate(R.layout.view_list_item, null);
    }
    //获取子item布局文件中的控件
    ImageView imgView = (ImageView)convertView.findViewById(R.id.lvew_img);
    TextView txtTitle = (TextView)convertView.findViewById(R.id.lvew_title);

    File file = getItem(position);        //根据position获取数据
    if(null != file)
    {
        imgView.setImageResource(file.isDirectory() ? R.drawable.folder : R.drawable.file);
        txtTitle.setText(file.getName());
    }
    return convertView;
```

}

FileMgr 代码：

```java
public class FileMgr {

    public List<File> getSubFiles(File file)
    {
        File[] files =  file.listFiles(new FilenameFilter(){
            @Override
            public boolean accept(File f, String name) {
                return !name.startsWith(".");
            }
        });
        System.out.println("files:" + (null == files));
        if(null != files)
        {
            System.out.println("file mgr length:" + files.length);
        }
        List<File> result = Arrays.asList(files);
        Collections.sort(result, new CustomFileComparator());
        return result;
    }
}
```

在 FileMgr 中，调用了 File 原生的 listFiles()方法获取文件的子文件集合，只不过通过文件名对隐藏文件做了过滤。在返回结果之前，使用自定义的 Comparator 对文件集合做了排序，自定义 Comparator 的代码如下：

```java
public class CustomFileComparator implements Comparator {

    @Override
    public int compare(Object o1, Object o2) {
        File file1 = (File)o1;
        File file2 = (File)o2;

        if(file1.isDirectory())
        {
            if(file2.isFile())
            {
                return -1;
```

```
            }
            return file1.getName().compareTo(file2.getName());
        }
        if(file2.isDirectory())
        {
            return 1;
        }
        return file1.getName().compareTo(file2.getName());
    }
}
```

第6章

ContentProvider

■ 对于一个 Android 设备来说，系统可能会包含多个应用程序。有些时候，不同的应用之间可能需要共享数据。例如手机的短信功能，在发信息的时候需要从通讯录中查找联系人，接收到陌生人的短信支持将陌生人的号码添加到通讯录中，短信应用和通讯录应用之间存在着数据共享。为了可以在不同的应用之间实现这种数据交换，Android 提供了 ContentProvider 接口，ContentProvider 也是 Android 的四大组件之一，当一个应用需要把自己的数据对外暴露时，可以实现自己的 ContentProvider，其他的应用可以通过 ContentResovler 访问 ContentProvider 提供的数据。Android 系统已经为一些常见的数据提供了标准的 ContentProvider，如联系人、视频、图片等。

6.1 ContentProvider 和 URI 简介

ContentProvider

无论数据的存储方式是什么，ContentProvider 以表的形式组织数据，并呈现给其他应用程序。例如 Android 系统内置的用户字典，它会存储用户想要保存的非标准字词的拼写，数据的组织形式如表 6.1 所示。每一行表示一条数据记录，每一列表示数据的一个字段。

表 6.1 用户字典的数据示例

单词	应用 ID	频率	语言区域	_ID
mapreduce	user1	100	en_US	1
precompiler	user14	200	fr_FR	2
applet	user2	225	fr_CA	3
const	user1	255	pt_BR	4
int	user5	100	en_UK	5

开发者如果想将自己应用程序中的数据暴露给其他应用，必须继承 ContentProvider 基类实现数据的增删改查。基类 ContentProvider 定义了相关的方法供子类实现，具体如表 6.2 所示。

表 6.2 ContentProvider 提供的方法

方法
abstract boolean onCreate() 其他应用程序第一次访问 ContentProvider 时会回调该方法，可以在其中做一些初始化操作
abstract Uri insert(Uri uri, ContentValues values) 子类实现该方法处理一条插入请求
abstract int delete(Uri uri, String selection, String[] selectionArgs) 子类实现该方法处理删除请求
abstract int update(Uri uri, ContentValues values, String selection, String[] selectionArgs) 子类实现该方法处理更新请求
abstract Cursor query(Uri uri, String[] projection, String selection, String[] selectionArgs, String sortOrder) 子类实现该方法处理查询请求
abstract String getType(Uri uri) 获取数据的 MIME 类型

每个应用程序都可以实现自己访问 ContentProvider 对外暴露数据，那其他应用程序该访问哪个 ContentProvider 呢？例如，短信应用想要读取通讯录中的联系人，怎样才能知道哪个是通讯录提供的 ContentProvider 呢？其实从表 6.2 所示的方法中也可以看出，ContentProvider 是通过统一资源定位符（URI）来标识自己的。Android 中 URI 的形式一般如下：content://authority/XXX。其中 content://是固定的格式，类似于网络请求中的 http://。authority 唯一标识了一个 ConentProvider，系统就是根据

authority 部分找到对应的 ContentProvider。XXX 部分指向了 ContentProvider 中的数据，这部分可以是动态改变的。另外，许多内容提供者都允许通过将 ID 值追加到 URI 的末尾来访问数据表中的某一行，例如对于 content://user_dictionary/words/4，user_dictionary 就是 URI 的 authority 部分，words/4 表示访问 words 数据的第 4 条记录。

6.2 创建 ContentProvider

一般在开发过程中，很少需要实现自定义的 ContentProvider，因为很少有应用程序需要对外暴露自己的数据。但是，了解 ContentProvider 的创建过程有利于我们更好地理解如何使用 ContentResovler 访问系统已经提供的 ContentProvider。实现自定义 ContentProvider 有以下几个步骤。

（1）定义一个名为 CONTENT_URI 的 URI 对象，用作该 ContentProvider 的标识。

（2）定义一个类继承自 ContentProvider 类，并实现父类的 onCreate()、insert()、delete()、update()、query() 和 getType() 方法。

（3）在 AndroidManifest.xml 中配置该 ContentProvider。

下面我们通过示例代码具体说明一下 ContentProvider 创建过程。

案例 6.1　创建 ContentProvider，对外提供学生信息

本例程序对外提供学生信息，包括学生的学号、姓名、年龄。

- 案例代码

（1）创建一个工具类用于保存 ContentProvider 需要用到的常量，示例代码如下：

```java
public class Constants implements BaseColumns {
    //URI的authority部分
    public static final String AUTHORITY = "com.demo.student.provider";
    //ContentProvider的CONTENT_URI
    public static final Uri CONTENT_URI = Uri.parse("content://" + AUTHORITY + "/students");
    //定义数据的MIME类型（多行数据）
    public static final String DATA_TYPE = "vnd.android.cursor.dir/vnd.com.demo.provider.student";
    //定义数据的MIME类型（单行数据）
    public static final String DATA_TYPE_ITEM = "vnd.android.cursor.item/vnd.com.demo.provider.student";
    //数据表的字段
    public static final String _ID = "_id";
    public static final String SNO = "sno";
    public static final String SNAME = "sname";
    public static final String SAGE = "sage";
}
```

Constants 类定义了一些 ContentProvider 需要用到的常量，首先定义了一个 URI 类型的变量 CONTENT_URI，用于标识当前的 ContentProvider。然后定义了两个 String 变量 DATA_TYPE 和 DATA_TYPE_ITEM，用于表示 ContentProvider 数据的 MIME 类型，其中 DATA_TYPE 代表多行数据的类型，DATA_TYPE_ITEM 代表单行数据，"vnd.android.cursor.dir"和"vnd.android.cursor.item"为固定写法。最后，定义了数据表的若干字段，其中_ID 是必需的，用于唯一标识一条数据记录。

（2）实现一个类继承自 ContentProvider 类，数据的存储方式可以有多种，示例代码采用 SQLite 进行数据的存取，具体如下：

```java
public class CustomProvider extends ContentProvider {
    private static final UriMatcher matcher = new UriMatcher(UriMatcher.NO_MATCH);
    private static final int STUDENT_ALL = 0;
    private static final int STUDNET_ITEM = 1;
    private DBOpenHelper mDBHelper;
    static
    {
        matcher.addURI(Constants.AUTHORITY, "students", STUDENT_ALL);
        matcher.addURI(Constants.AUTHORITY, "students/#", STUDNET_ITEM);
    }
    @Override
    public boolean onCreate() {
        mDBHelper = new DBOpenHelper(getContext(), "students.db", 1);
        return true;
    }

    @Override
    public int delete(Uri uri, String selection, String[] selectionArgs) {
        SQLiteDatabase db = mDBHelper.getWritableDatabase();
        int num = 0;
        switch(matcher.match(uri))      //解析URI
        {
        case STUDENT_ALL:       //更新多条记录
        {
            num = db.delete("student", selection, selectionArgs);
            break;
        }
        case STUDNET_ITEM:      //删除单条记录
        {
```

```java
            long id = ContentUris.parseId(uri);         //解析URI中的id
            String str = Constants._ID + "=" + id;      //根据id组成查询条件
            if(!TextUtils.isEmpty(selection))
            {
                //加上传递过来的查询条件
                selection = selection + " and " + str;
            }
            num = db.delete("student", selection, selectionArgs);
            break;
        }
        default:
            num = 0;
            break;
    }
    //通知数据发生改变
    getContext().getContentResolver().notifyChange(uri, null);
    return num;
}


@Override
public Uri insert(Uri uri, ContentValues values) {
    SQLiteDatabase db = mDBHelper.getWritableDatabase();
    //向数据库中插入一条数据
    long rowId = db.insert("student", Constants._ID, values);
    if(rowId > 0)       //如果插入成功
    {
        //向数据表的URI追加新行的Id值
        Uri newRowUri = ContentUris.withAppendedId(uri, rowId);
        //通知数据发生改变
        getContext().getContentResolver().notifyChange(newRowUri, null);
        return newRowUri;      //返回新行数据的URI
    }
    return null;
}


@Override
public Cursor query(Uri uri, String[] projection, String selection,
```

```java
                String[] selectionArgs, String sortOrder) {
    SQLiteDatabase db = mDBHelper.getReadableDatabase();
    switch(matcher.match(uri))       //解析URI
    {
        case STUDENT_ALL:       //访问多条记录
        {
            return db.query("student", projection, selection, selectionArgs, null, null, sortOrder);
        }
        case STUDNET_ITEM:      //访问单条记录
        {
            long id = ContentUris.parseId(uri);        //从URI中解析出id
            String str = Constants._ID + "=" + id;     //根据id组成查询条件
            if(!TextUtils.isEmpty(selection))
            {
                //加上传递过来的查询条件
                selection = selection + " and " + str;
            }
            return db.query("student", projection, selection, selectionArgs, null, null, sortOrder);
        }
        default:
            return null;
    }
}

@Override
public int update(Uri uri, ContentValues values, String selection,
        String[] selectionArgs) {
    SQLiteDatabase db = mDBHelper.getWritableDatabase();
    int num = 0;
    switch(matcher.match(uri))       //解析URI
    {
        case STUDENT_ALL:       //更新多条记录
        {
            num = db.update("student", values, selection, selectionArgs);
            break;
        }
        case STUDNET_ITEM:      //更新单条记录
```

```
            {
                long id = ContentUris.parseId(uri);        //解析URI中的id
                String str = Constants._ID + "=" + id;     //根据id组成查询条件
                if(!TextUtils.isEmpty(selection))
                {
                    //加上传递过来的查询条件
                    selection = selection + " and " + str;
                }
                num = db.update("student", values, selection, selectionArgs);
                break;
            }
        default:
            num = 0;
            break;
        }
        //通知数据发生改变
        getContext().getContentResolver().notifyChange(uri, null);
        return num;
    }


    @Override
    public String getType(Uri uri) {
        switch (matcher.match(uri)) {
        case STUDENT_ALL:        //多行数据
            return Constants.DATA_TYPE;
        case STUDNET_ITEM:       //单行数据
            return Constants.DATA_TYPE_ITEM;
        default:
            return null;
        }
    }
}
```

　　在自定义的 ContentProvider 中有一个很重要的类 UriMatcher，用来注册当前 ContentProvider 可以匹配的 URI，注册是通过 addURI(String authority, String path, int code)将一个 URI 对象和一个标识码对应起来。例如示例中的 matcher.addURI(Constants.AUTHORITY, "students", STUDENT_ALL)就将"content://com.demo.student.provider/students" 和标识码 0 对应起来。除了注册可以匹配的 URI 之外，还需要重写父类的若干方法。其中，可以在 onCreate()方法中做一些初始化的操作。insert()方法用来插

入数据,这里直接向数据库中插入一条数据即可。delete()、update()、query()方法分别用来删除、更新、查询数据,其操作方式比较类似,首先调用 UriMatcher 的 match(Uri uri)方法判断 URI 的类型,如果是操作多条数据,直接调用数据库的相关方法,如果是操作单条数据,首先调用 ContentUris 的 parseId(Uri uri)方法解析出需要操作的数据 id 号,然后组成查询条件操作指定行数的数据。getType()方法用于返回指定 URI 对象的数据类型,根据多行数据还是单行数据返回预先定义好的字符串。

(3) 在 AndroidManifest.xml 中配置,具体如下:

```
<provider
        android:name="com.demo.contentprovider.CustomProvider"
        android:authorities="com.demo.student.provider"/>
```

6.3 使用 ContentResovler 操作数据

使用 ContentResovler 类可以访问别的应用程序通过 ContentProvider 提供的数据,常见的方法如表 6.3 所示。

表 6.3 ContentResovler 类常见方法

方法名	方法说明
insert(Uri url, ContentValues values)	向 URI 对应的 ContentProvider 中插入数据
delete(Uri url, String where, String[] selectionArgs)	删除数据
update(Uri uri, ContentValues values, String where, String[] selectionArgs)	更新数据
query(Uri uri, String[] projection, String selection, String[] selectionArgs, String sortOrder)	查询数据

Android 系统提供了很多标准的 ContentProvider 可供访问,如日历程序、联系人、照相机等,对应的 CONTENT_URI 在 android.provider 中。下面我们以访问联系人数据为例,说明一下 ContentResovler 的使用方法。

案例 6.2 使用 ContentResovler 添加、查询联系人

本例在 Activity 中定义两个按钮,一个用于添加联系人,一个用于查询联系人。

- **案例代码**

Activity 代码:

```
public class MainActivity extends Activity implements OnClickListener{
    protected void onCreate(Bundle savedInstanceState) {
        super.onCreate(savedInstanceState);
        setContentView(R.layout.activity_main);
```

```java
        Button btnAdd = (Button)findViewById(R.id.btn_add);         //获取按钮控件
        Button btnQuery = (Button)findViewById(R.id.btn_query);
        btnAdd.setOnClickListener(this);        //设置监听器
        btnQuery.setOnClickListener(this);
    }
    public void onClick(View v) {       //处理按钮单击事件
        switch(v.getId())
        {
        case R.id.btn_add:      //添加
        {
            add();
            break;
        }
        case R.id.btn_query:        //查询
        {
            query();
            break;
        }
        default:
            break;
        }
    }
    private void add()
    {
        ContentValues values = new ContentValues();
        //先插入一条空数据,获取当前通讯录表中的id
        Uri rawContactUri = getContentResolver().insert(RawContacts.CONTENT_URI, values);
        long rawContactId = ContentUris.parseId(rawContactUri);
        //添加联系人的姓名
        values.put(Data.MIMETYPE, StructuredName.CONTENT_ITEM_TYPE);        //插入单条数据
        values.put(Data.RAW_CONTACT_ID, rawContactId);      //设置id
        values.put(StructuredName.GIVEN_NAME, "zhangsan");      //设置名称
        getContentResolver().insert(ContactsContract.Data.CONTENT_URI, values);     //插入
        //添加联系人的电话号码
        values.put(Data.MIMETYPE, Phone.CONTENT_ITEM_TYPE);
        values.put(Data.RAW_CONTACT_ID, rawContactId);      //设置id
        values.put(Phone.NUMBER, "13812345678");        //设置电话号码
```

```
            values.put(Phone.TYPE, Phone.TYPE_MOBILE);    //设置电话号码的类型
            getContentResolver().insert(ContactsContract.Data.CONTENT_URI, values);
    }
    private void query()
    {
            Cursor cursor = getContentResolver().query(ContactsContract.Contacts.CONTENT_URI, null,
null, null, null);     //查询联系人的数据
            while(cursor.moveToNext())       //遍历
            {
                //获取数据的id
                String id=cursor.getString(cursor.getColumnIndex(ContactsContract.Contacts._ID));
                String        //获取联系人的显示姓名
    name=cursor.getString(cursor.getColumnIndex(ContactsContract.Contacts.DISPLAY_NAME));
                Cursor      //根据id查询联系人的电话号码
    c=getContentResolver().query(ContactsContract.CommonDataKinds.Phone.CONTENT_URI, null,
    ContactsContract.CommonDataKinds.Phone._ID +"=" + id, null, null);
                while(c.moveToNext())     //遍历取出联系人的电话号码
                {
                    String phone=c.getString(c.getColumnIndex(ContactsContract.
    CommonDataKinds.Phone.NUMBER));
                    Log.i("contact", "name: " + name + "; phone: " + phone);
                }
            }
    }
}
```

只要根据对应数据的 URI，就可以调用 ContentResovler 类的增删改查方法进行相应的处理，ContentResovler 对象可以通过调用 getContentResolver()方法获得。

6.4　本章小结

ContentProvider 也是 Android 的四大组件之一，它为系统内不同应用程序之间的数据交换提供了一种标准接口，一些应用程序可以通过这些标准将内部的数据暴露出去供其他应用访问。大部分情况下，应用都是使用其他应用暴露的接口，但是为了更好地使用这些接口，本章在开始的部分我们介绍了如何创建 ContentProvider，最后介绍了如何通过 ContentResolver 使用其他应用程序和系统提供的数据。

6.5 小练习

下面我们通过一个小练习来了解一下 ContentResolver 的使用场景和示例,本练习会访问系统中所有的联系人并将其用列表展示出来,如图 6.1 所示。长按联系人姓名会弹出菜单执行相关操作,如图 6.2 所示。单击"删除"按钮会删除当前联系人,单击"拨号"按钮会拨打该号码。

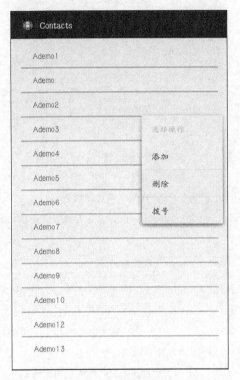

图 6.1　联系人列表　　　　图 6.2　操作菜单

MainActivity 代码:

```
public class MainActivity extends Activity {
    private ListViewAdapter mAdapter;
    @Override
    protected void onCreate(Bundle savedInstanceState) {
        super.onCreate(savedInstanceState);
        setContentView(R.layout.activity_main);
        mAdapter = new ListViewAdapter(this);
        init();
        ListView listView = (ListView)findViewById(R.id.list_view);
        listView.setAdapter(mAdapter);
        listView.setOnCreateContextMenuListener(new OnCreateContextMenuListener(){
```

```java
            @Override
            public void onCreateContextMenu(ContextMenu menu, View v,
                    ContextMenuInfo menuInfo) {
                menu.setHeaderTitle("选择操作");
                menu.add(0, 0, 0, "添加");
                menu.add(0, 1, 0, "删除");
                menu.add(0, 2, 0, "拨号");
            }
        });
    }

    private void init()
    {
        List<Contact> contacts = new ArrayList<Contact>();
        Cursor cursor = getContentResolver().query(ContactsContract.Contacts.CONTENT_URI, null, null, null, null);    //查询联系人的数据
        while(cursor.moveToNext()) //遍历
        {
            //获取数据的id
            String id = cursor.getString(cursor.getColumnIndex(ContactsContract.Contacts._ID));
            //获取联系人的显示姓名
            String name = cursor.getString(cursor.getColumnIndex(ContactsContract.Contacts.DISPLAY_NAME));
            //根据Id查询联系人的电话号码
            Cursor c = getContentResolver().query(ContactsContract.CommonDataKinds.Phone.CONTENT_URI, null, ContactsContract.CommonDataKinds.Phone._ID +"=" + id, null, null);
            List<String> phones = new ArrayList<String>();
            while(c.moveToNext())      //遍历取出联系人的电话号码
            {
                String phone=c.getString(c.getColumnIndex(ContactsContract.CommonDataKinds.Phone.NUMBER));
                phones.add(phone);
            }
            contacts.add(new Contact(id, name, phones));
        }
        mAdapter.updateContacts(contacts);
    }
```

```java
@Override
public boolean onContextItemSelected(MenuItem item) {
    AdapterContextMenuInfo menuInfo = (AdapterContextMenuInfo) item.getMenuInfo();
    //info.id得到listview中选择的条目绑定的id
    int position = menuInfo.position;
    Contact contact = mAdapter.getItem(position);
    switch (item.getItemId()) {
    case 0:
        Toast.makeText(this, "add:" + contact.getContactName(), Toast.LENGTH_SHORT).show();
        return true;
    case 1:
        deleteContact(contact);
        return true;
    case 2:
        Intent intent = new Intent(Intent.ACTION_CALL);
        Uri data = Uri.parse("tel:" + contact.getPhone().get(0));
        intent.setData(data);
        startActivity(intent);
        return true;
    default:
        return super.onContextItemSelected(item);
    }
}

private void deleteContact(Contact contact)
{
    ContentResolver resolver = getContentResolver();
    //删除Contacts表中的数据
    resolver.delete(ContactsContract.Contacts.CONTENT_URI, ContactsContract.Contacts._ID + " =?", new String[]{String.valueOf(contact.getContactId())});
    //删除RawContacts表中的数据
    resolver.delete(ContactsContract.RawContacts.CONTENT_URI, ContactsContract.RawContacts.CONTACT_ID + " =?", new String[]{String.valueOf(contact.getContactId())});
    //删除姓名
    resolver.delete(RawContacts.CONTENT_URI, "display_name=?", new String[]
```

```
{contact.getContactName()});
        }
    }
```

在 MainActivity 的 init()方法中,程序根据 getContentResolver()获取系统的联系人 id、姓名和电话号码,设置到 listView 中列表展示。同时通过 setOnCreateContextMenuListener()方法为 listView 关联上下文菜单,当在 listview 的 item 上长按时,会弹出上下文菜单。单击"添加"按钮,会弹出 Toast 提示当前选择的联系人姓名。单击"删除"按钮,会根据 id 删除联系人的电话号码,同时删除联系人的姓名信息。单击"拨号"按钮时,会通过 Intent 调用 Intent.ACTION_CALL 的系统 action,从而实现拨号功能。

该实例需要在 AndroidManifest.xml 文件中注册联系人访问权限和拨号权限,具体如下:

```xml
<uses-permission android:name="android.permission.GET_ACCOUNTS"/>
<uses-permission android:name="android.permission.READ_CONTACTS"/>
<uses-permission android:name="android.permission.WRITE_CONTACTS"/>
<uses-permission android:name="android.permission.CALL_PHONE" />
```

第7章

Service

■ Service 也是 Android 的四大组件之一，是一个可以在后台长时间运行的组件。Service 可以由其他组件启动，而且即使用户切换到其他应用，Service 仍将在后台运行。Service 的功能和使用方法与 Activity 比较类似，区别在于 Activity 需要向用户提供一个界面供交互，而 Service 在后台执行一些操作，如处理一些网络下载请求、后台播放音乐、文件读写等。

7.1 Service 简介

Service

7.1.1 创建、配置 Service

Service 的创建步骤和 Activity 一致，自定义的 Service 需要继承基类 Service，并且需要实现基类的若干方法，其中必须实现的方法是 onBind()方法，一些比较常见且重要的方法如表 7.1 所示。

表 7.1 Service 中的常见方法

方法名	方法说明
IBinder onBind(Intent intent)	该方法返回一个 IBinder 对象，用于其他组件和 Service 通信，子类必须实现该方法
void onCreate()	Service 第一次创建时会回调该方法
int onStartCommand()	每次调用 startService(Intent)启动 Service 时都会回调该方法
boolean onUnbind(Intent intent)	当 Service 上绑定的所有组件都解绑时会回调该方法
void onDestroy()	Service 被销毁时会回调该方法

Service 创建示例：

```
public class DemoService extends Service {
    @Override
    public IBinder onBind(Intent intent) {
        // TODO Auto-generated method stub
        return null;
    }
    @Override
    public void onCreate() {
        // TODO Auto-generated method stub
        super.onCreate();
    }
    @Override
    public void onDestroy() {
        // TODO Auto-generated method stub
        super.onDestroy();
    }
    @Override
    public int onStartCommand(Intent intent, int flags, int startId) {
```

```
        // TODO Auto-generated method stub
        return super.onStartCommand(intent, flags, startId);
    }
    @Override
    public boolean onUnbind(Intent intent) {
        // TODO Auto-generated method stub
        return super.onUnbind(intent);
    }
}
```

和 Activity 一样，Service 在创建完成之后，还需要在 AndroidManifest.xml 文件中对 Service 进行配置，否则系统找不到定义的 Service。Service 的配置方法是在 manifest 文件的<application/>元素中增加一个<service/>元素，代码示例如下，其中 android:name 属性指定了 DemoService 的位置。另外，Service 也可以配置<intent-filter/>属性，用于指定 Service 可以匹配的 action。

Service 配置示例：

```xml
<service android:name="com.demo.service.DemoService">
        <intent-filter >
            <action android:name="com.demo.demoservice"/>
        </intent-filter>
</service>
```

Service 在创建和配置完成之后，我们就可以在程序中运行该 Service 了。Service 的运行方式有两种：启动 Service 和绑定 Service。下面我们将分别介绍这两种使用方式。

7.1.2 启动和停止 Service

启动 Service 通过调用 startService()方法实现，一旦 Service 被启动，可在后台无限期运行，即使启动该 Service 的组件已经被销毁。这种方式的 Service 通常用于执行比较单一且不需要将执行结果返回给调用者的任务。例如，网络下载或上传文件，操作完成之后会自行停止 Service。下面我们通过一个实例说明一下在 Activity 中启动 Service 的方法。

案例 7.1　启动和停止 Service

本例 Activity 中包含两个按钮，一个按钮用于启动 Service，另一个按钮用于停止 Service。

- 案例代码

MainActivity 代码：

```java
public class MainActivity extends Activity {
    protected void onCreate(Bundle savedInstanceState) {
        super.onCreate(savedInstanceState);
        setContentView(R.layout.activity_main);
```

```java
            initWidget();
    }
    private void initWidget()
    {
        Button btnStart = (Button)findViewById(R.id.btn_start);    //获取"启动"按钮
        Button btnStop  = (Button)findViewById(R.id.btn_stop);     //获取"停止"按钮
        btnStart.setOnClickListener(new OnClickListener() {
            public void onClick(View v) {
                Intent intent = new Intent(MainActivity.this, DemoService.class);
                startService(intent);      //调用startService启动Service
            }
        });
        btnStop.setOnClickListener(new OnClickListener() {
            public void onClick(View v) {
                Intent intent = new Intent(MainActivity.this, DemoService.class);
                stopService(intent);       //调用stopService停止Service
            }
        });
    }
}
```

DemoService 代码：

```java
public class DemoService extends Service {
    @Override
    public IBinder onBind(Intent intent) {
        Log.i("DemoService", "onBind()");
        return null;
    }
    @Override
    public void onCreate() {
        Log.i("DemoService", "onCreate()");
        super.onCreate();
    }
    @Override
    public void onDestroy() {
        Log.i("DemoService", "onDestroy()");
        super.onDestroy();
    }
```

```
    @Override
    public int onStartCommand(Intent intent, int flags, int startId) {
        Log.i("DemoService", "onStartCommand()");
        return super.onStartCommand(intent, flags, startId);
    }
    @Override
    public boolean onUnbind(Intent intent) {
        Log.i("DemoService", "onUnbind()");
        return super.onUnbind(intent);
    }
}
```

- **案例分析**

在 DemoService 的每个方法中没有做具体的操作，只是打印了一条语句。单击 Activity 中的"启动"按钮，控制台日志打印如图 7.1 所示，回调了 Service 中的 onCreate()和 onStartCommand()方法。

```
I  06-30 02:33:50.772   25752   25752   com.demo.service   DemoService   onCreate()
I  06-30 02:33:50.772   25752   25752   com.demo.service   DemoService   onStartCommand()
```

图 7.1　启动 Service

在 Service 已经启动的情况下单击"启动"按钮再次调用 startService()，控制台打印如图 7.2 所示，说明每次调用 startService 方法都会回调 onStartCommand()方法。

```
I  06-30 02:33:50.772   25752   25752   com.demo.service   DemoService   onCreate()
I  06-30 02:33:50.772   25752   25752   com.demo.service   DemoService   onStartCommand()
I  06-30 02:34:59.260   25752   25752   com.demo.service   DemoService   onStartCommand()
```

图 7.2　再次调用 startService()

单击 Activity 中的"停止"按钮调用 stopService()方法，控制台打印如图 7.3 所示，Service 已经停止，回调了 onDestroy()方法。

```
I  06-30 02:33:50.772   25752   25752   com.demo.service   DemoService   onCreate()
I  06-30 02:33:50.772   25752   25752   com.demo.service   DemoService   onStartCommand()
I  06-30 02:34:59.260   25752   25752   com.demo.service   DemoService   onStartCommand()
I  06-30 02:35:58.930   25752   25752   com.demo.service   DemoService   onDestroy()
```

图 7.3　停止 Service

7.1.3　绑定 Service

调用 startService()之后，运行的 Service 和调用者之间基本不存在太大的关联。但有些时候需要允许 Service 和其他组件之间进行数据交换，例如手机音乐播放器在后台播放音乐，在 Activity 页面上需要知道当前播放的歌曲状态。这种情况可以通过将一个组件和 Service 进行绑定，绑定 Service 通过调用 bindService(Intent service, ServiceConnection conn, int flags)实现，该方法包含 3 个参数，第 1 个参数 Service 指定需要绑定的 Service，第 3 个参数 flags 指定绑定时是否自动创建 Service，第 2 个参数是一个 ServiceConnection 对象，用于监听其他组件和 Service 之间的连接情况，不能为空，定义的方式如下：

```java
private ServiceConnection sc = new ServiceConnection()
{
    public void onServiceConnected(ComponentName name, IBinder service) {
    }
    public void onServiceDisconnected(ComponentName name) {
    }
};
```

当组件和 Service 连接成功之后，会回调 onServiceConnected()方法，当组件和 Service 断开连接之后，会回调 onServiceDisconnected()方法。onServiceConnected()方法中包含一个 IBinder 对象，组件和 Service 之间的交互就通过这个 IBinder 对象实现。我们在 7.1.1 小节介绍 Service 的常用方法时说过，创建自定义的 Service 必须实现基类的 IBinder onBind(Intent intent)方法，当组件和 Service 绑定后，该方法返回的 IBinder 对象会传到 ServiceConnection 对象中作为 onServiceConnected()方法的参数，从而实现 Service 和其他组件之间的通信。IBinder 在 API 中是一个接口，其中的一个实现类是 Binder 类。在开发中，可以通过继承 Binder 类实现自己的 IBinder 对象。下面我们通过一个具体的实例说明一个 Service 的绑定和解绑。

案例 7.2　绑定和解绑 Service

本例仍然是在 Activity 中实现绑定 Service、获取信息和解绑 Service。

- **案例代码**

MainActivity 代码：

```java
public class MainActivity extends Activity implements OnClickListener {
    private DemoService.MyBinder mBinder = null;      //定义一个Binder对象用于消息交互
private ServiceConnection sc = new ServiceConnection()      //定义一个ServiceConnection对象
    {
            //Activity绑定Service成功
        public void onServiceConnected(ComponentName name, IBinder service) {
            Log.i("DemoService", "=========service connected========");
            if(null != service)
            {
                //将传递过来的IBinder对象赋值
```

```java
                mBinder = (DemoService.MyBinder)service;
            }
        }

        public void onServiceDisconnected(ComponentName name) {
            Log.i("DemoService", "=========service disconnected========");
        }
    };
    protected void onCreate(Bundle savedInstanceState) {
        super.onCreate(savedInstanceState);
        setContentView(R.layout.activity_main);
        initWidget();
    }
    private void initWidget()
    {
        Button btnStart  = (Button)findViewById(R.id.btn_bind);      //获取"绑定"按钮
        Button btnGetMsg = (Button)findViewById(R.id.btn_getmsg);    //单击用于获取消息
        Button btnStop   = (Button)findViewById(R.id.btn_unbind);    //获取"解绑"按钮
        btnStart.setOnClickListener(this);
        btnGetMsg.setOnClickListener(this);
        btnStop.setOnClickListener(this);
    }
    public void onClick(View v) {
        switch(v.getId())
        {
        case R.id.btn_bind:      //"绑定"按钮
        {
            Intent intent = new Intent(MainActivity.this, DemoService.class);
            bindService(intent, sc, Service.BIND_AUTO_CREATE);       //绑定Service
            break;
        }
        case R.id.btn_getmsg:    //获取消息按钮
        {
            if(null != mBinder)
            {
                //通过Binder对象获取Service的数据
                Log.i("DemoService", mBinder.getMsg());
```

```
                }
                break;
            }
            case R.id.btn_unbind:        //"解绑"按钮
            {
                unbindService(sc);       //解绑Service
                break;
            }
            default:
                break;
        }
    }
}
```

DemoService 代码：

```
public class DemoService extends Service {
    private String strMsg = "this is a message from DemoService !";
    private MyBinder mBinder = new MyBinder();
    public class MyBinder extends Binder         //定义一个类继承自Binder类
    {
        public String getMsg()           //对外提供访问Service中数据的方法
        {
            return strMsg;
        }
    }
    public IBinder onBind(Intent intent) {
        Log.i("DemoService", "onBind()");
        return mBinder;              //返回一个Binder对象
    }
    @Override
    public void onCreate() {
        Log.i("DemoService", "onCreate()");
        super.onCreate();
    }
    @Override
    public void onDestroy() {
        Log.i("DemoService", "onDestroy()");
        super.onDestroy();
```

```
    }
    @Override
    public int onStartCommand(Intent intent, int flags, int startId) {
        Log.i("DemoService", "onStartCommand()");
        return super.onStartCommand(intent, flags, startId);
    }
    @Override
    public boolean onUnbind(Intent intent) {
        Log.i("DemoService", "onUnbind()");
        return super.onUnbind(intent);
    }
}
```

- **案例分析**

本例在 Activity 中定义了 3 个按钮，分别用于绑定 Service、获取 Service 中数据、解绑 Service。DemoService 中继承 Binder 类定义了一个内部类 MyBinder，用于对外提供方法访问 Service 中的数据，onBind()方法返回 MyBinder 类的一个对象。单击"绑定"按钮，控制台打印结果如图 7.4 所示，DemoService 回调了 onCreate()和 onBind()方法，Activity 和 Service 连接成功之后，ServiceConnection 对象回调了 onServiceConnected()方法，Activity 在该方法中获取 MyBinder 对象。

```
I  06-30 06:29:25.750  18269  18269  com.demo.service  DemoService  onCreate()
I  06-30 06:29:25.750  18269  18269  com.demo.service  DemoService  onBind()
I  06-30 06:29:25.788  18269  18269  com.demo.service  DemoService  =========service connected=========
```

图 7.4 绑定 Service

单击"获取数据"按钮，控制台打印结果如图 7.5 所示，说明 Service 中的消息正确地传递到了 Activity 中。

```
I  06-30 06:29:25.750  18269  18269  com.demo.service  DemoService  onCreate()
I  06-30 06:29:25.750  18269  18269  com.demo.service  DemoService  onBind()
I  06-30 06:29:25.788  18269  18269  com.demo.service  DemoService  =========service connected=========
I  06-30 06:30:05.231  18269  18269  com.demo.service  DemoService  this is a message from DemoService !
```

图 7.5 通过 Binder 对象传递消息

单击 Activity 中的"解绑"按钮，控制台打印结果如图 7.6 所示，Service 回调了 onBind()和 onDestroy()方法。

```
I  06-30 06:29:25.750  18269  18269  com.demo.service  DemoService  onCreate()
I  06-30 06:29:25.750  18269  18269  com.demo.service  DemoService  onBind()
I  06-30 06:29:25.788  18269  18269  com.demo.service  DemoService  =========service connected=========
I  06-30 06:30:05.231  18269  18269  com.demo.service  DemoService  this is a message from DemoService !
I  06-30 06:30:56.060  18269  18269  com.demo.service  DemoService  onUnbind()
I  06-30 06:30:56.060  18269  18269  com.demo.service  DemoService  onDestroy()
```

图 7.6 解绑 Service

7.2　Service 的生命周期

Service 的生命周期比较简单，但是对于不同的运行方式，Service 的生命周期也有所不同。如图 7.7 所示，左图是启动 Service 的生命周期，右图是绑定 Service 的生命周期。

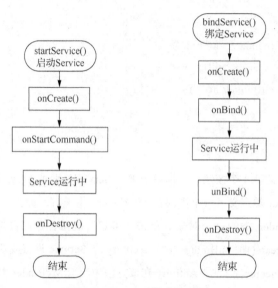

图 7.7　Service 的生命周期

从图 7.7 中可以看出，无论是启动 Service 还是绑定 Service，生命周期都从回调 onCreate()开始，到回调 onDestroy()结束。onCreate()方法中可以做一些初始化的工作，onDestroy()中可以做资源的释放。例如，后台播放音乐可以在 onCreate()方法中做播放器和音频文件的初始化操作，可以在 onDestory()中做播放器资源的释放。回调 onCreate()方法之后，对于启动 Service 会执行 onStartCommand()方法。而对于绑定 Service，会回调 onBind()方法，解绑 Service 之后，会回调 unBind()方法。

7.3　跨进程调用 Service

在 7.2 节中我们介绍了 Service 的使用方法，但这些都是在同一个应用程序中启动 Service。在 Android 上，不同的进程之间通常也会需要互相访问数据，这就需要实现跨进程通信。如果两个进程之间需要通信，则需要通信接口，Android 使用 AIDL 来定义进程之间的通信接口。

7.3.1　创建 Service 和 AIDL 接口

跨进程绑定 Service 与之前在同一个应用内调用 bindService()类似，onBind()方法返回一个 IBinder 对象，在 ServiceConnection 对象的 onServiceConnection()方法中获取 IBinder 对象，通过 IBinder 对象

实现通信。只不过跨进程时，需要通过 AIDL 处理 IBinder 对象的传递和获取。AIDL 的定义和 Java 接口的定义类似，区别在于 AIDL 文件以.aidl 结尾，示例代码如下：

```
package com.demo.aidl.service;

interface IStu
{
    String getStuNo();
    String getStuName();
}
```

在 src 目录下创建以.aidl 结尾的文件之后，SDK 工具会在项目的 gen/目录中生成 IBinder 接口文件。生成的文件名与.aidl 文件名一致，只是使用了.java 扩展名，如图 7.8 所示。

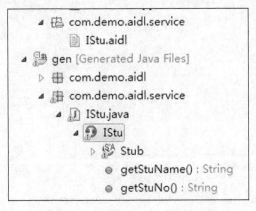

图 7.8　gen 目录下生成的与 AIDL 文件同名的 Java 文件

打开自动生成的 IStu.java 文件可以看到，里面定义了一个 Stub 内部类，该内部类继承自 Binder 类并实现了 IStu.aidl 接口，因此，该类的对象可以作为 Service 中 onBind()函数的返回值。

AIDL 接口创建好之后，就可以创建 Service 类，Service 类的创建和之前类似，具体代码如下：

```
public class AIDLService extends Service {

    private String stuNo;
    private String stuName;

    private IStu.Stub mBinder = new IStu.Stub() {     //定义一个IStu.Stub类的对象
        public String getStuNo() throws RemoteException {      //实现对应的方法
            return stuNo;
        }
        public String getStuName() throws RemoteException {
            return stuName;
```

```java
        }
    };
    public IBinder onBind(Intent intent) {
        return mBinder;        //返回IBinder类型的变量
    }
    @Override
    public void onCreate() {
        super.onCreate();
        stuNo = "201701";

    }
    @Override
    public void onDestroy() {
        stuName = "Tom";
        super.onDestroy();
    }
}
```

AIDLService 的配置：

```xml
<service android:name="com.demo.aidl.service.AIDLService">
    <intent-filter >
        <action android:name="com.demo.aidlservice"/>
    </intent-filter>
</service>
```

可以看到，用于跨进程通信的 Service 与之前一样，只是增加了 AIDL 接口，IBinder 对象为生成的 Stub 类的对象。

7.3.2　跨进程绑定 Service

Service 定义好之后，就可以在别的进程中对该 Service 进行访问，绑定的方式和之前一样，只是对 Binder 对象的获取方式有所不同，具体代码如下。

MainActivity 代码：

```java
package com.demo.aidl;

import com.demo.aidl.service.IStu;

import android.os.Bundle;
import android.os.IBinder;
```

```java
import android.os.RemoteException;
import android.app.Activity;
import android.app.Service;
import android.content.ComponentName;
import android.content.Intent;
import android.content.ServiceConnection;
import android.util.Log;
import android.view.Menu;
import android.view.View;
import android.view.View.OnClickListener;
import android.widget.Button;

public class MainActivity extends Activity implements OnClickListener {
    private IStu mBinder;
    private ServiceConnection sc = new ServiceConnection()        //定义ServiceConnection对象
    {
        public void onServiceConnected(ComponentName name, IBinder service) {
            Log.i("DemoService", "=========service connected========");
            if(null != service)
            {
                mBinder = IStu.Stub.asInterface(service);        //获取传递过来的Binder对象
            }
        }

        public void onServiceDisconnected(ComponentName name) {
            Log.i("DemoService", "=========service disconnected========");
        }
    };
    protected void onCreate(Bundle savedInstanceState) {
        super.onCreate(savedInstanceState);
        setContentView(R.layout.activity_main);
        initWidget();
    }
    private void initWidget()
    {
        Button btnStart = (Button)findViewById(R.id.btn_bind);
        Button btnGetMsg = (Button)findViewById(R.id.btn_getmsg);
```

```java
        Button btnStop  = (Button)findViewById(R.id.btn_unbind);
        btnStart.setOnClickListener(this);
        btnGetMsg.setOnClickListener(this);
        btnStop.setOnClickListener(this);
    }
    public void onClick(View v) {
        switch(v.getId())
        {
        case R.id.btn_bind:
            {
                Intent intent = new Intent();
                intent.setAction("com.demo.aidlservice");
                bindService(intent, sc, Service.BIND_AUTO_CREATE);        //绑定Service
                break;
            }
        case R.id.btn_getmsg:
            {
                if(null != mBinder)
                {
                    try {
                        Log.i("DemoService", "stuNo:" + mBinder.getStuNo() + "; stuName:" + mBinder.getStuName());        //通过Binder对象访问其他进程中Service的数据
                    } catch (RemoteException e) {
                        e.printStackTrace();
                    }
                }
                break;
            }
        case R.id.btn_unbind:
            {
                unbindService(sc);        //解绑Service
                break;
            }
        default:
            break;
        }
```

        }
    }

可以看到，跨进程访问 Service 的代码和之前一样，只不过需要注意以下两点。

（1）Binder 对象的类型为 IStu 类型，这个数据类型是在 Service 进程中定义 AIDL 接口之后自动生成的，当前进程中不存在该类型。这里只需要把 Service 进程中的 AIDL 文件复制到当前项目的 src 文件夹下，当前项目也会自动在 gen 目录下生成对应的 IStu.java 文件。

（2）在 ServiceConnection 对象回调 onServiceConnected(ComponentName name, IBinder service)之后，之前是直接通过强制类型转换将参数 service 转换为自定义的 Binder 对象。这里是通过调用 asInterface(service)方法将参数 service 转换为 IStu 对象。

7.4 本章小结

本章我们主要介绍了 Android 的 Service 组件，Service 和同为 Android 四大组件的 Activity 比较类似，只不过 Service 运行在后台，没有界面显示，常用于一些服务和监控程序。和介绍 Activity 一样，我们首先介绍了 Service 的创建、启动、停止和绑定，然后介绍了 Service 的生命周期，有助于读者更好地了解 Service 从创建到消亡的整个过程。最后介绍了 Service 的跨进程调用。

7.5 小练习

下面我们通过一个小练习简单了解一下 Service 的应用场景。本练习结合第 4 章的广播接收者实现了一个短信监听程序，程序启动一个 Service，常驻后台，其中注册一个自定义的广播监听者，当系统接收到短信消息时，自定义的广播接收者会拦截读取短信的发送者和短信内容，具体实现如下。

MainActivity 代码：

```java
public class MainActivity extends Activity {
    @Override
    protected void onCreate(Bundle savedInstanceState) {
        super.onCreate(savedInstanceState);
        setContentView(R.layout.activity_main);
        Intent intent = new Intent(MainActivity.this, MessageService.class);
        startService(intent);
    }
}
```

MainActivity 的代码比较简单，仅仅是通过 Intent 启动了一个 MesageService。

MessageService 代码：

```java
public class MessageService extends Service {

    private MessageReciever mReceiver;

    @Override
    public IBinder onBind(Intent arg0) {
        return null;
    }

    @Override
    public void onCreate() {
        mReceiver = new MessageReciever();
        IntentFilter filter = new IntentFilter("android.provider.Telephony.SMS_RECEIVED");
        registerReceiver(mReceiver, filter);
    }

    @Override
    public void onDestroy() {
        if(null != mReceiver)
        {
            unregisterReceiver(mReceiver);
            mReceiver = null;
        }
    }

    @Override
    public int onStartCommand(Intent intent, int flags, int startId) {
        return super.onStartCommand(intent, flags, startId);
    }

    @Override
    public boolean onUnbind(Intent intent) {
        return super.onUnbind(intent);
    }
}
```

MessageService 常驻后台运行,在 onCreate()方法中,动态注册了一个自定义的广播接收者,接收 Intent 为 android.provider.Telephony.SMS_RECEIVED 的消息。广播接收者的代码如下:

```java
public class MessageReciever extends BroadcastReceiver{

    private static final String SMS_RECEIVER_ACTION = "android.provider.Telephony.SMS_RECEIVED";
    @Override
    public void onReceive(Context context, Intent intent) {
        StringBuilder sBuilder = new StringBuilder();
        if(SMS_RECEIVER_ACTION.equals(intent.getAction()))
        {
            Bundle bundle = intent.getExtras();
            if(null != bundle)
            {
                Object[] pdus = (Object[])bundle.get("pdus");
                SmsMessage[] messages = new SmsMessage[pdus.length];
                for(int i = 0; i < messages.length; ++i)
                {
                    messages[i] = SmsMessage.createFromPdu((byte[])pdus[i]);
                }
                for(SmsMessage msg : messages)
                {
                    sBuilder.append("source:").append(msg.getDisplayOriginatingAddress()).append("message content:").append("\n");
                    sBuilder.append(msg.getDisplayMessageBody()).append("\n");
                }
            }
        }
        Toast.makeText(context, "receive messages:\n" + sBuilder.toString(), Toast.LENGTH_LONG).show();
    }
}
```

在 MessageReciever 中,首先判断接收到消息的 intent 是否为短信消息,然后调用 Intent.getExtras() 获取广播携带的消息,最后解析短信的发送者和短信内容,以 Toast 的方式显示出来。

示例需要短消息的接收权限,所以需要在 AndroidManifest.xml 文件中声明,具体如下:

```
<!-- 发送消息-->
```

```
<uses-permission android:name="android.permission.SEND_SMS"/>
<!-- 阅读消息 -->
<uses-permission android:name="android.permission.READ_SMS"/>
<!-- 写入消息 -->
<uses-permission android:name="android.permission.WRITE_SMS" />
<!-- 接收消息 -->
<uses-permission android:name="android.permission.RECEIVE_SMS"/>
```

第8章

高级编程

■ 前面几章我们介绍了 Android 开发的一些基础知识，本章我们将会在此基础上做一些拓展，如网络编程。对于一个 Android 设备来说，网络请求是很常见的操作，所以掌握一些网络编程的知识是必要的。另外，在 UI 方面，Android 系统提供了一些动画的 API，可以提高用户体验。在与用户的交互中，如果一个操作比较耗时，可能会造成用户长时间等待的现象，甚至会因为 ANR（Application Not Responding）造成应用异常退出。本章我们会介绍如何使用线程做耗时的操作，另外还会介绍 Fragment 和 RecycleView 组件。

8.1 网络编程

网络编程

网络编程是指多个设备之间通过网络进行数据交换，网络通信基于"请求-响应模型"，即一台设备发送通信请求，另一台设备进行反馈。发送请求端称为客户端，响应请求端称为服务端。例如常见的 QQ 程序，用户打开 QQ 客户端程序之后，输入账号和密码，再单击"登录"，即向腾讯服务端发送登录请求，服务端把请求结果反馈到客户端。Android 是基于 Java 进行开发的，所以 JDK 中关于网络编程的 API 在 Android 中均可使用。

8.1.1 TCP、UDP 协议基础

两台设备之间进行通信，一定要有通信协议，即客户端以一定的格式将数据发送出去，服务端接收到数据之后，可以根据同样的协议将数据的内容解析出来。在现有的网络中，通信方式有两种：TCP 和 UDP。TCP 是传输控制协议，提供的是面向连接、可靠的字节流服务，通信双方必须先建立一个 TCP 连接，然后才能传输数据。而且 TCP 协议还提供了超时重发、数据校验、拥塞控制等功能，保证了数据的可靠传输。UDP 是用户数据报协议，只是简单地把数据报发送出去，但是并不能保证数据能到达目的地，不提供可靠性。所以 TCP 应用于需要安全可靠传输数据的场景，但是开销比较大。UDP 应用于对数据可靠性要求不是太高的传输，其优点是开销小，另外 UDP 没有数据校验、拥塞控制等操作，故而传输比较快。

对于上层应用来说，无论是 TCP 还是 UDP，通信都包括客户端和服务端，处理流程也具有一般性，具体如表 8.1 所示。

表 8.1 客户端和服务端的处理流程

客户端处理流程： （1）根据服务器的 IP 地址和端口号建立网络连接 （2）建立连接之后，进行数据交换，向服务端发送请求和接收服务端反馈的数据 （3）关闭连接
服务端处理流程： （1）服务端启动之后，监听一个固定的端口，被动地等待客户端连接 （2）客户端连接到服务端之后，服务端可以获取客户端的 IP 地址等信息，可以进行数据交换 （3）接收客户端发送的数据，然后把处理的结果反馈给客户端 （4）关闭连接

8.1.2 Socket 通信

对于 TCP 和 UDP 的通信，Java 均有对应的 API。现在我们以 TCP 为例，通过一个实例说明网络通信的方式。

案例 8.1　实现网络通信

对于 TCP 通信，Java 中使用 Socket 类做客户端开发，使用 ServerSocket 做服务端开发。

- **案例代码**

客户端代码：

```java
private void startClient() {
    String data = "hello World";
    InputStream is = null;
    OutputStream os = null;
    Socket c = null;
    try {
        c = new Socket("192.168.1.101", 7005);    //建立连接
        os = c.getOutputStream();         //获得输出流
        os.write(data.getBytes());        //发送数据

        is = c.getInputStream();          //获得输入流
        byte[] b = new byte[1024];
        is.read(b);        //接收反馈数据
    } catch (Exception e) {
        e.printStackTrace();
    } finally {
        try {
            if (null != c) {
                c.close();
            }
            if(null != is)
            {
                is.close();
                is = null;
            }
            if(null != os)
            {
                os.close();
                os = null;
            }
        } catch (IOException e) {
            e.printStackTrace();
```

```
            }
        }
    }
```

代码分析：客户端首先定义了一个 Socket 对象向服务端请求建立连接，构造函数以服务器的 IP 地址和端口号作为参数。建立连接之后，调用 getOutputStream()获取输出流，向服务端发送数据。

服务端代码：

```
private void startServer() {
        ServerSocket sc = null;
        try
        {
            sc = new ServerSocket(7005);        //监听端口号
            while(true)
            {
                Socket c = sc.accept();         //被动地等待连接
                new ServerThread(c);            //开启一个线程处理客户端请求
            }
        }catch(Exception e)
        {
            e.printStackTrace();
        }finally
        {
            try
            {
                if(null != sc)
                {
                    sc.close();
                    sc = null;
                }
            }catch(Exception e1)
            {
                e1.printStackTrace();
            }
        }
    }
```

代码分析：服务端首先定义了一个 ServerSocket 对象，构造函数以将要监听的端口号作为参数，然后调用 accept()方法被动地等待连接。这个方法是一个阻塞方法，如果没有收到客户端请求，代码会阻塞在该处，不会向下执行。如果收到客户端请求，则会返回一个与客户端之间的 **Socket** 连接，然后开启一个线程处理客户端的请求（因为一个服务端可能同时对应多个客户端，所以为每个客户端开一个线程处理）。**ServerThread** 代码包含的是基本的输入/输出流操作，具体如下。

ServerThread 代码：

```java
public class ServerThread extends Thread {
    private Socket c = null;
    public ServerThread(Socket c)
    {
        this.c = c;
        start();         //启动线程
    }
    @Override
    public void run() {
        InputStream is = null;
        try
        {
            if(null != c)
            {
                is = c.getInputStream();      //获取输入流
            }
            if(null != is)
            {
                byte[] b = new byte[1024];
                int size = is.read(b);        //读取接收到的消息
                Log.i("Socket", "the msg is:" + new String(b, 0, size));
            }
        }catch(Exception e)
        {
            e.printStackTrace();
        }finally
        {
            try
            {
                if(null != is)
                {
                    is.close();
                    is = null;
```

```
                }
                if(null != c)
                {
                    c.close();
                    c = null;
                }
            }catch(Exception e1)
            {
                e1.printStackTrace();
            }
        }
    }
}
```

8.1.3 下载网络资源

在 Android 开发中,经常需要请求网络资源,例如播放在线音乐,或者加载显示一张网络图片。Java 提供了 HttpURLConnection 和 HttpsURLConnection,两者都可以基于 URL 实现简单的请求响应功能,区别在于是访问 http 链接还是访问 https 链接。

案例 8.2 下载网络图片

本例通过 HttpURLConnection 获取网络图片。

- **案例代码**

```
private Bitmap getBitmap(String path)
    {
        Bitmap bm = null;
        try
        {
            URL url = new URL(path);          //创建一个URL对象,参数为网络图片的链接地址
            //使用URL对象开启一个连接
            HttpURLConnection con = (HttpURLConnection)url.openConnection();
            //设置相关参数
            con.setDoInput(true);
            con.setConnectTimeout(5000);
            con.setReadTimeout(2000);
            con.connect();
            InputStream is = con.getInputStream();     //获取输入流
            bm = BitmapFactory.decodeStream(is);        //将输入流解码为Bitmap对象
```

```
                is.close();
        }catch(Exception e)
        {
                e.printStackTrace();
        }
        return bm;
}
```

- **案例分析**

代码中首先根据网络图片的链接地址创建了一个 URL 对象，然后调用 openConnection()方法开启一个连接，接着设置相关参数，如超时时间等。然后调用 getInputStream()方法获取输入流，后续就是输入/输出流的基本处理，可以调用 BitmapFactory 的 decodeStream()方法将输入流解析成 Bitmap 图片。

8.2 图形图像和动画

对于一个应用来说，图片是一种很丰富的表达形式。Android 中也为图片的处理提供了大量的 API，不仅包括图片的显示、绘制，还包括一些简单的动画效果。本节我们将介绍这些 API 的使用方法。

图形图像和动画

8.2.1 Bitmap 和 BitmapFactory

Android 中提供了 Bitmap 类用于图片处理，一个 Bitmap 对象代表一张位图，存储了图片的宽高、颜色、像素点等信息，Bitmap 类提供了大量的方法，其常见的方法如表 8.2 所示。

表 8.2　Bitmap 常见的方法

static Bitmap createBitmap(Bitmap source, int x, int y, int width, int height) 静态方法，以 source 图片的(x,y)位置为起点，截取宽为 width、高为 height 的图片
static Bitmap createBitmap(int width, int height, Bitmap.Config config) 创建一个 Bitmap 对象
static Bitmap createScaledBitmap(Bitmap src, int dstWidth, int dstHeight, boolean filter) 将 src 图像缩放后创建一个新的 Bitmap 对象
final int getHeight() 获取 Bitmap 的高
final int getWidth() 获取 Bitmap 的宽
void recycle() 回收 Bitmap 对象和对应像素点所占的内存

续表

final boolean isRecycled()
判断 Bitmap 对象是否被回收
int getPixel(int x, int y)
获取图片指定位置处像素点的值
void setPixel(int x, int y, int color)
设置图片指定位置处像素点的值

BitmapFactory 主要用于加载 Bitmap 对象,可以从资源文件加载,也可以根据图片的路径进行加载,还可以根据输入流对 Bitmap 对象进行解析,相应的方法如表 8.3 所示。

表 8.3　BitmapFactory 的方法

static Bitmap decodeByteArray(byte[] data, int offset, int length, BitmapFactory.Options opts)
将字节数组解码成 Bitmap 对象
static Bitmap decodeFile(String pathName, BitmapFactory.Options opts)
根据图片的路径加载 Bitmap 对象
static Bitmap decodeResource(Resources res, int id, BitmapFactory.Options opts)
将资源文件解析成 Bitmap 对象
static Bitmap decodeStream(InputStream is)
将输入流解析成 Bitmap 对象

8.2.2　Android 绘图基础

除了显示已有的图片之外,Android 还支持一些简单的二维绘图,其实对于 Android 的一些基本组件,如 TextView、Button 等,也都是系统绘制出来的,绘制的操作在 View 类的 onDraw(Canvas canvas) 方法中,每个组件需要实现 onDraw(Canvas canvas) 方法进行自定义的绘制。所以,Android 的绘图应该定义一个类继承自 View 组件,并重新定义 onDraw(Canvas canvas) 方法。其中,参数 Canvas 可以理解为画布,绘制操作均在 Canvas 上执行。Canvas 支持的一些方法如表 8.4 所示。

表 8.4　Canvas 类支持的操作举例

drawBitmap(Bitmap bitmap, float left, float top, Paint paint)
从 Bitmap 对象的左上角开始绘制
drawCircle(float cx, float cy, float radius, Paint paint)
绘制一个圆

续表

drawLine(float startX, float startY, float stopX, float stopY, Paint paint) 绘制一条线
drawPoint(float x, float y, Paint paint) 绘制一个点
drawRect(float left, float top, float right, float bottom, Paint paint) 绘制一个矩形
drawText(String text, float x, float y, Paint paint) 绘制一个字符串

从表 8.4 中可以看到，每一个方法中都包含一个 Paint 类的参数，Paint 类代表画笔，它指定了画笔的颜色和粗细等，相关的方法如表 8.5 所示。

表 8.5 Paint 类相关的方法

setAlpha(int a) 设置画笔的透明度
setAntiAlias(boolean aa) 设置是否抗锯齿
setColor(int color) 设置画笔的颜色
setShader(Shader shader) 设置画笔的填充效果
setShadowLayer(float radius, float dx, float dy, int color) 设置画笔的阴影效果
setStrokeWidth(float width) 设置画笔的粗细
setTextSize(float textSize) 设置绘制的文字大小

案例 8.3 使用线性布局

本例说明绘制 API 的方法，运行结果如图 8.1 所示。

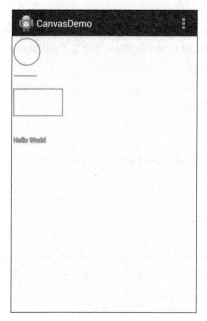

图 8.1　图形的绘制

- **案例代码**

自定义 View 代码：

```
public class CanvasView extends View {        //自定义一个类继承自View组件
    public CanvasView(Context context, AttributeSet attrs) {
        super(context, attrs);
    }
    @Override
    protected void onDraw(Canvas canvas) {        //重写onDraw()方法
        super.onDraw(canvas);
        Paint paint = new Paint();       //定义一个画笔对象
        paint.setAntiAlias(true);        //设置抗锯齿
        paint.setStyle(Paint.Style.STROKE);      //设置画笔风格
        paint.setStrokeWidth(5);        //设置画笔粗细
        paint.setColor(Color.GREEN);        //设置颜色
        paint.setTextSize(24);       //设置文字显示大小
        canvas.drawCircle(60, 60, 50, paint);        //绘制圆
        canvas.drawLine(10, 100, 100, 100, paint);       //绘制线
        canvas.drawRect(10, 150, 100, 300, paint);       //绘制矩形
        canvas.drawText("Hello World", 10, 300, paint);      //绘制文字
    }
}
```

8.2.3 补间动画

在 Android 应用中，经常会出现一些动画效果，例如控件的滑入滑出、图片的渐隐等。常见的实现方式有补间动画和属性动画，补间动画是指开发者指定好控件的初始状态和结束状态，系统自动补齐显示控件的中间状态。Android 补间动画支持的效果比较简单，包括平移、缩放、旋转、透明度变化，对应的类如表 8.6 所示。

表 8.6 补间动画对应的类

类	对应的动画
TranslateAnimation	用于做平移动画的类，需要指定控件的起始和结束时的位置
ScaleAnimation	用于做缩放动画的类，需要指定动画的缩放中心、起始时的缩放比和结束时的缩放比
RotateAnimation	用于做旋转动画的类，需要指定旋转中心的坐标、起始的旋转角度和结束时的旋转角度
AlphaAnimation	用于做透明度变化的类，需要指定起始的透明度和结束时的透明度

下面我们通过一个实例说明一下补间动画的使用。

案例 8.4　使用补间动画

本例在 Activity 上显示一张图片和 4 个按钮，4 个按钮分别用于触发 4 种动画，运行结果如图 8.2 所示。

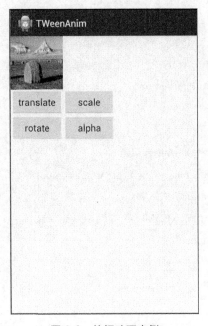

图 8.2　补间动画实例

- 案例代码

MainActivity 代码:

```java
public class MainActivity extends Activity implements OnClickListener {
    protected void onCreate(Bundle savedInstanceState) {
        super.onCreate(savedInstanceState);
        setContentView(R.layout.activity_main);
        initWidget();
    }
    private void initWidget() {
        Button btnTrans = (Button) findViewById(R.id.btn_translate);    //获取对应的4个按钮
        Button btnScale = (Button) findViewById(R.id.btn_scale);
        Button btnRotate = (Button) findViewById(R.id.btn_rotate);
        Button btnAlpha = (Button) findViewById(R.id.btn_alpha);
        btnTrans.setOnClickListener(this);    //设置监听事件
        btnScale.setOnClickListener(this);
        btnRotate.setOnClickListener(this);
        btnAlpha.setOnClickListener(this);
    }
    public void onClick(View v) {
        ImageView imgView = (ImageView) findViewById(R.id.img_pic);
        switch (v.getId()) {
        case R.id.btn_translate: {    //平移动画
            TranslateAnimation tanim = new TranslateAnimation(0, 100, 0, 0);
            tanim.setDuration(500);    //设置动画执行的时间
            tanim.setFillAfter(true);    //设置动画执行后是否保持变化后的状态
            imgView.startAnimation(tanim);
            break;
        }
        case R.id.btn_scale: {    //缩放动画
            ScaleAnimation sanim = new ScaleAnimation(0.0f, 1.2f, 0.0f, 1.2f,
                    Animation.RELATIVE_TO_SELF, 0.5f,
                    Animation.RELATIVE_TO_SELF, 0.5f);
            sanim.setDuration(500);
            sanim.setFillAfter(true);
            imgView.startAnimation(sanim);
            break;
        }
```

```java
            case R.id.btn_rotate: {        //旋转动画
                RotateAnimation ranim = new RotateAnimation(0, 360,
                        Animation.RELATIVE_TO_SELF, 0.5f,
                        Animation.RELATIVE_TO_SELF, 0.5f);
                ranim.setDuration(500);
                ranim.setFillAfter(true);
                imgView.startAnimation(ranim);
                break;
            }
            case R.id.btn_alpha: {        //透明度变化
                AlphaAnimation anim = new AlphaAnimation(1.0f, 0.0f);
                anim.setDuration(500);
                anim.setFillAfter(true);
                imgView.startAnimation(anim);
                break;
            }
            default:
                break;
        }
    }
```

8.2.4 属性动画

属性动画是在 API 11 之后加入的功能，它几乎可以作用在任何对象上，而且不同于补间动画只能支持 4 种变换，属性动画是在一定时间内将对象的属性从一个初始值改变到另一个值，因此，只要是对象存在的属性，无论是可见还是不可见的，都可以实现动画效果。属性动画可以通过 ObjectAnimator 实现。下面我们通过一个实例说明一下属性动画的具体属性。

> **案例 8.5　使用属性动画**

本例在 Activity 中定义了两个按钮，一个用于触发图片的平移，另一个用于触发图片的透明度的变换。

- 案例代码

MainActivity 代码：

```java
public class MainActivity extends Activity implements OnClickListener {
    protected void onCreate(Bundle savedInstanceState) {
        super.onCreate(savedInstanceState);
        setContentView(R.layout.activity_main);
```

```
            initWidget();
    }
    private void initWidget() {
        Button btnTrans = (Button) findViewById(R.id.btn_translation);    //获取按钮
        Button btnAlpha = (Button) findViewById(R.id.btn_alpha);
        btnTrans.setOnClickListener(this);    //设置监听事件
        btnAlpha.setOnClickListener(this);
    }
    @SuppressLint("NewApi")
    public void onClick(View v) {
        ImageView imgView = (ImageView) findViewById(R.id.img_pic);
        switch (v.getId()) {
        case R.id.btn_translation: {    //平移
            ObjectAnimator anim = ObjectAnimator.ofFloat(imgView, "translationX", 0, 100);
            anim.setDuration(500);
            anim.start();
            break;
        }
        case R.id.btn_alpha: {    //透明度变换
            ObjectAnimator anim = ObjectAnimator.ofFloat(imgView, "alpha", 1.0f, 0.0f);
            anim.setDuration(500);
            anim.start();
            break;
        }
        default:
            break;
        }
    }
}
```

- **案例分析**

单击按钮后，使用 ObjectAnimator 类实现属性动画，首先调用 ofFloat()方法获取 ObjectAnimator 对象，ofFloat()方法第 1 个参数为需要变换的对象，第 2 个参数为需要改变的属性，"translationX" 表示水平方向上的平移，"alpha" 为透明度的改变，其他的还有 "rotationX" "rotationY" 等，设置好需要控制的对象和属性之后，调用 setDuration()设置动画执行的时间，最后调用 start()方法执行动画。

8.3 多媒体应用开发

多媒体应用开发

在 Android 应用中，经常需要播放媒体资源，如音乐播放器、视频播放器和游戏的背景音乐等。Android 提供了 MediaPlayer 类，可以很简单地实现播放本地存储或者网络上的音/视频文件。

8.3.1 MediaPlayer 类介绍

MediaPlayer 提供了大量的方法可以控制音/视频的播放、暂停、定位等，常见的方法如表 8.7 所示。

表 8.7 MediaPlayer 类的方法

方法名	方法说明
int getCurrentPosition()	获取当前播放的位置
int getDuration()	获取音/视频文件的总长
int getVideoHeight()	获取视频的高度
int getVideoWidth()	获取视频的宽度
boolean isPlaying()	判断当前是否正在播放
Void pause()	暂停播放
void prepare()	MediaPlayer 开始准备（同步方法）
void prepareAsync()	MediaPlayer 开始准备（异步方法）
void release()	释放 MediaPlayer 占用的资源
void reset()	重置 MediaPlayer 的状态
void seekTo(int msec)	定位到音/视频的指定位置，可用于实现快进、快退
void setDataSource(String path)	设置视频源
setDataSource(Context context, Uri uri)	设置视频源
setDisplay(SurfaceHolder sh)	用于显示播放画面
setOnCompletionListener(MediaPlayer.OnCompletionListener listener)	设置视频播放完成之后的监听事件
setOnErrorListener(MediaPlayer.OnErrorListener listener)	设置播放出现错误时的监听事件
setOnPreparedListener(MediaPlayer.OnPreparedListener listener)	设置 MediaPlayer 准备好时的监听事件
setOnSeekCompleteListener(MediaPlayer.OnSeekCompleteListener listener)	设置定位完成时的监听事件

续表

方法名	方法说明
void start()	开始播放
void stop()	停止播放

使用 MediaPlayer 播放音/视频时,首先调用 setDataSource()方法设置需要播放的视频源,然后调用 prepare()或 prepareAsync()方法做视频播放的准备工作。两个方法的区别在于 prepareAsync()是异步方法,不会阻塞程序的执行,可以通过调用 setOnPreparedListener()方法监听播放器是否已经准备好,准备好之后就可以调用 start()方法进行播放。

8.3.2 使用 MediaPlayer 和 SurfaceView 播放视频

播放的视频必须显示在 View 上,而且视频的画面一直在改变,所以 View 需要一直重绘。对于这种需要不断更新 View 内容的场景,Android 提供了 SurfaceView 类,SurfaceView 关联了一个 SurfaceHolder 对象,专门用于绘制 SurfaceView 的内容。SurfaceHolder 会有 3 个回调方法反馈 SurfaceView 的状态,具体如表 8.8 所示。

表 8.8 SurfaceHolder 的回调方法

方法	说明
surfaceChanged(SurfaceHolder holder, int format, int width, int height)	SurfaceView 的大小发生改变
surfaceCreated(SurfaceHolder holder)	SurfaceView 第一次被创建完成
surfaceDestroyed(SurfaceHolder holder)	SurfaceView 被销毁时回调该方法

下面我们通过一个具体的实例说明使用 MediaPlayer 和 SurfaceView 播放视频的方法。

案例 8.6 使用 MediaPlayer 和 SurfaceView 播放视频

本例实现视频播放,单击"暂停"按钮时,暂停播放视频;单击"播放"按钮时,重新开始播放视频。最终可以退出播放页面。

- 案例代码

MainActivity 代码:

```java
public class MainActivity extends Activity implements OnClickListener {

    private SurfaceView mSurface;
    private Button mBtnPlay;
    private Button mBtnPause;
```

```java
private MediaPlayer mPlayer = null;

protected void onCreate(Bundle savedInstanceState) {
    super.onCreate(savedInstanceState);
    setContentView(R.layout.activity_main);
    initWidget();
    initPlayer();
}

private void initWidget() {
    mSurface = (SurfaceView) findViewById(R.id.surfaceView);        //获取SurfaceView控件
    mBtnPlay = (Button) findViewById(R.id.btn_play);
    mBtnPause = (Button) findViewById(R.id.btn_pause);
    mBtnPlay.setOnClickListener(this);
    mBtnPause.setOnClickListener(this);
}

public void onClick(View v) {
    switch (v.getId()) {
    case R.id.btn_play: {        //单击"播放"按钮
        if(null != mPlayer && !mPlayer.isPlaying())        //判断当前是否已经处于播放状态
        {
            mPlayer.start();        //播放
        }
        break;
    }
    case R.id.btn_pause: {        //单击"暂停"按钮
        if(null != mPlayer && mPlayer.isPlaying())        //判断当前是否处于播放状态
        {
            mPlayer.pause();        //暂停
        }
        break;
    }
    default:
        break;
    }
```

```java
}
private void initPlayer()
{
    mPlayer = new MediaPlayer();         //定义一个MediaPlayer对象
    //获取SurfaceView关联的SurfaceHolder
    SurfaceHolder holder = mSurface.getHolder();
    holder.addCallback(surfaceCallBack);    //为SurfaceHolder添加回调函数
    try {
        mPlayer.setOnPreparedListener(onPreparedListener);    //设置监听事件
        mPlayer.setDataSource("mnt/sdcard/demo.avi");    //设置视频源
        mPlayer.prepareAsync();     //调用异步准备方法
    } catch (Exception e) {
        e.printStackTrace();
    }
}

private OnPreparedListener onPreparedListener = new OnPreparedListener()
{
    public void onPrepared(MediaPlayer mp) {    //MediaPlayer准备完成，开始播放
        mp.start();
    }
};

private SurfaceHolder.Callback surfaceCallBack = new SurfaceHolder.Callback() {
    public void surfaceDestroyed(SurfaceHolder holder) {
        // TODO Auto-generated method stub
    }
    public void surfaceCreated(SurfaceHolder holder) {    //SurfaceView创建成功
        mPlayer.setDisplay(holder);    //为MediaPlayer设置SurfaceHolder
    }

    @Override
    public void surfaceChanged(SurfaceHolder holder, int format, int width,
            int height) {
        // TODO Auto-generated method stub
    }
```

```
    };
    protected void onDestroy() {      //页面销毁时释放MediaPlayer占用的资源
        if(null != mPlayer)
        {
            mPlayer.stop();
            mPlayer.release();
            mPlayer = null;
        }
        super.onDestroy();
    }
}
```

act_main.xml 布局文件：

```xml
<LinearLayout xmlns:android="http://schemas.android.com/apk/res/android"
    xmlns:tools="http://schemas.android.com/tools"
    android:layout_width="match_parent"
    android:layout_height="match_parent"
    android:layout_gravity="center"
    android:orientation="vertical">
    //定义一个SurfaceView布局
    <SurfaceView
        android:id="@+id/surfaceView"
        android:layout_width="match_parent"
        android:layout_height="400dp"
        />
    <LinearLayout
        android:layout_width="match_parent"
        android:layout_height="wrap_content"
        android:layout_gravity="center_horizontal"
        android:orientation="horizontal">
        <Button
            android:id="@+id/btn_play"
            android:layout_width="100dp"
            android:layout_height="100dp"
            android:gravity="center"
            android:textSize="18sp"
            android:text="play"/>
```

```xml
            <Button
                android:id="@+id/btn_pause"
                android:layout_width="100dp"
                android:layout_height="100dp"
                android:gravity="center"
                android:textSize="18sp"
                android:text="pause"/>
    </LinearLayout>
</LinearLayout>
```

- 案例分析

实例中首先获取了 SurfaceView 对象，然后调用 getHolder()方法获取关联的 SurfaceHolder 对象，设置监听事件，当 SurfaceView 创建完成之后将 SurfaceHolder 设置给 MediaPlayer。然后调用 setDataSource()方法设置视频源，为 MediaPlayer 设置监听事件。接着调用 prepareAsync()方法做 MediaPlayer 的准备工作，准备完成之后会回调 onPrepared()方法，在其中开始播放视频。单击"暂停"按钮时，调用 pause()方法暂停播放视频；单击"播放"按钮时，调用 play()方法重新开始播放视频。退出播放页面时，调用 release()方法释放 MediaPlayer 占用的资源。

8.4 线程开发

线程开发

线程在程序开发中是一个很重要的概念，在 Android 中，线程分为主线程和子线程。主线程又叫作 UI 线程，主要处理和界面有关的事情，用于界面的绘制和交互。对于用户来说，随时都有可能操作页面，而且对响应速度要求较高，因此，主线程中不能做太耗时的操作，否则会给用户视觉上造成卡顿的现象，甚至有可能会因为执行阻塞产生 ANR（Application Not Responding）而导致应用异常退出。

Android 基于 Java 开发，所以对于线程也有很好的支持，除了 JDK 中支持的一些 API 之外，Android 也针对自身的机制做了一些拓展。下面我们将介绍两个常用的类：AsyncTask 和 ThreadPoolExecutor。

8.4.1 AsyncTask 及其使用

AsyncTask 是一个执行异步操作的类，它在主线程中创建和触发，但是在子线程中执行后台任务，然后将执行的进度和最终结果传递给主线程并在主线程中更新 UI。例如有一种很常见的操作，应用中如果要在 ImageView 控件中显示一张网络图片，首先需要从网络上下载图片，然后显示到控件上。图片的下载依赖于网络情况，如果在主线程中执行下载操作，可能会造成用户长时间等待，并且不能及时响应用户的操作。这时就可以使用 AsyncTask 类在子线程中执行下载操作，然后通知主线程下载完成并显示。AsyncTask 有 3 个主要的回调方法，如表 8.9 所示。

表 8.9　AsyncTask 的主要回调方法

方法名	方法说明
doInBackground()	在后台执行任务
onProgressUpdate()	回调当前执行的进度
onPostExecute()	任务执行完成

具体使用方法如下所示：

```java
public void onClick(View v) {
    new DownloadImageTask().execute("http://example.com/image.png");         //执行异步任务
}
private class DownloadImageTask extends AsyncTask<String, Void, Bitmap>
    {
        protected Bitmap doInBackground(String... arg0) {    //在后台执行任务
            String url = args[0];    //获取参数
            Bitmap bitmap = downloadImgByUrl(url);    //执行下载任务
            return bitmap;    //返回执行结果
        }
        protected void onPostExecute(Bitmap result) {    //下载完成之后回调该方法
            mImageView.setImageBitmap(result);    //对返回结果做处理
        }
        protected void onProgressUpdate(Void... values) {    //回调下载进度
            super.onProgressUpdate(values);
        }
    }
```

上面代码中定义的一个内部类 DownloadImageTask 继承自 AsyncTask，并根据需要重写 AsyncTask 的方法。在 doInBackground() 方法中获取传递的参数，并调用方法执行图片下载操作。下载过程中会回调 onProgressUpdate() 方法返回进度。下载完成之后会回调 onPostExecute() 方法，在其中将返回的 Bitmap 显示到 ImageView 中。在主线程中，调用 execute() 方法执行异步任务，将图片的 URL 链接作为参数传入。

8.4.2　ThreadPoolExecutor 介绍

当一个应用中需要创建多个线程时，可以将其放入线程池中进行管理，Java 中使用 Executor 做线程池的管理和线程的调度。Executor 是一个接口，它的实现类为 ThreadPoolExecutor。创建 ThreadPoolExecutor 对象时，可以向构造函数传入一系列的参数来配置线程池。常用的构造函数有 ThreadPoolExecutor(int corePoolSize, int maximumPoolSize, long keepAliveTime, TimeUnit unit, BlockingQueue<Runnable> workQueue, ThreadFactory threadFactory)。其中，corePoolSize 是指线程池的

核心线程数，maximumPoolSize 指线程池可以容纳的最大的线程个数，keepAliveTime 指非核心线程在空闲时可以存活的时间，TimeUnit 是参数 keepAliveTime 的单位，workQueue 是指线程队列，ThreadFactory 指线程工厂，用于线程池创建线程。为了帮助读者理解以上各个参数的含义，下面我们说明一下线程池的工作步骤。

（1）当调用了 ThreadPoolExecutor 的 execute()方法要求执行一个任务时，线程池会判断当前的线程个数是否超过核心线程数，如果没有超过，则启动一个核心线程来执行任务。

（2）如果当前的线程个数超过了核心线程的个数，则将任务放到线程队列中排队等候。

（3）如果线程队列已满，则启动一个非核心线程来执行任务。

（4）如果线程的总个数已经达到了线程池最大的线程个数，则拒绝接收任务。

（5）如果线程池有线程处于空闲状态，则在指定 keepAliveTime 后回收线程，当线程个数等于核心线程数后，停止回收线程。

下面我们通过一个具体实例说明一个线程池的使用。

线程池的创建代码：

```
public class Main {
    public static void main(String[] args) {
        ThreadPoolExecutor executor = new ThreadPoolExecutor(1, 3, 1, TimeUnit.MINUTES, new ArrayBlockingQueue(4));    //创建一个线程池对象
        executor.execute(new Task());        //执行一个任务
    }
}
```

需要执行的任务：

```
public class Task extends Thread {        //继承Thread类
    public void run() {        //重写父类的run()方法
        super.run();
        System.out.println("this is a task in thread");        //执行自己的操作
    }
}
```

8.5　Fragment

为了更加动态和灵活地支持 UI 设计，Android 在 API 11 版本之后引入了 Fragment，它可以将 UI 碎片化，也可以被复用。Fragment 必须显示在 Activity 中，可以被动态地添加、移除、替换，但是每个 Fragment 也都具有自己的生命周期方法，并且可以各自处理用户的输入事件。常见的微信界面就可以使用 Fragment 来实现，如图 8.3 所示，单击下面的"微信"，上面就会对应显示聊天记录，单击"通讯录"，上面就会显示联系人列表。如果使用 Activity 实现比较

麻烦，但是如果使用 Fragment 实现就比较简单，界面的整体定义为一个 Activity，中间的内容区域设计为一个 Fragment，可以根据用户单击的 Item 动态地更换显示区域的 Fragment。

图 8.3　微信界面

8.5.1　Fragment 的创建

和 Activity 类似，创建自定义的 Fragment 需要继承自 Fragment 类，并实现父类的相关回调方法。其中比较常见的是 onCreate()、onCreateView()、onPause()方法。onCreate()方法在创建 Fragment 时会回调；onCreateView()在绘制 Fragment 视图的时候会回调，开发者需要在该方法中加载 Fragment 需要显示的布局文件；onPause()方法在用户离开 Fragment 时会回调。另外，Android 通过 FragmentManager 类管理在 Activity 中的 Fragment，具体的操作在 FragmentTransaction 中。首先可以通过 getFragmentManager()方法获取 FragmentManager 对象，然后调用 FragmentManager 的 beginTransaction()开启一个事务执行具体的操作。FragmentTransaction 类支持的常见操作如表 8.10 所示。

表 8.10　FragmentTransaction 类支持的操作

方法名	方法说明
add (int containerViewId, Fragment fragment)	在 Activity 指定的位置添加一个 Fragment
commit ()	提交事务
hide (Fragment fragment)	隐藏当前的 Fragment
remove (Fragment fragment)	从 Activity 中移除一个 Fragment
replace (int containerViewId, Fragment fragment)	在指定位置处替换一个 Fragment
show (Fragment fragment)	显示之前隐藏的 Fragment

下面我们通过一个具体的实例说明一下 Fragment 的用法，类似于微信，实例实现了单击底部按钮，

上部分内容区域动态改变的功能。

案例 8.7　单击底部按钮，上面内容区域动态改变

本例运行结果如图 8.4 所示，单击下面的按钮，上面的内容会动态改变。

图 8.4　Fragment 实例

- **案例代码**

FragmentFriend 代码：

```java
public class FragmentFriend extends Fragment {      //定义一个类继承自Fragment
    public void onCreate(Bundle savedInstanceState) {      //重写onCreate()方法
        super.onCreate(savedInstanceState);
    }
    //重写onCreateView()方法，在其中加载布局文件
    public View onCreateView(LayoutInflater inflater, ViewGroup container,
            Bundle savedInstanceState) {
        View view = inflater.inflate(R.layout.vew_fragment, container, false);     //加载布局文件
        TextView txtView = (TextView)view.findViewById(R.id.txt_title);
        txtView.setText("This is Friend UI");      //设置布局中的文字
        return view;
    }
}
```

FragmentContact 代码：

```java
public class FragmentContact extends Fragment {      //和FragmentFriend类代码类似
    public void onCreate(Bundle savedInstanceState) {
        super.onCreate(savedInstanceState);
```

```java
    }
    public View onCreateView(LayoutInflater inflater, ViewGroup container,
            Bundle savedInstanceState) {
        View view = inflater.inflate(R.layout.vew_fragment, container, false);
        TextView txtView = (TextView)view.findViewById(R.id.txt_title);
        txtView.setText("This is Contact UI");
        return view;
    }
}
```

view_fragment.xml 代码:

```xml
<?xml version="1.0" encoding="utf-8"?>
<LinearLayout xmlns:android="http://schemas.android.com/apk/res/android"
    android:layout_width="match_parent"
    android:layout_height="match_parent"
    android:orientation="vertical"
    android:gravity="center">
    <TextView
        android:id="@+id/txt_title"
        android:layout_width="300dp"
        android:layout_height="100dp"
        android:textSize="36sp"
        android:text="This is Friend UI"/>
</LinearLayout>
```

MainActivity 代码:

```java
public class MainActivity extends Activity implements OnClickListener {

    private Fragment mFriendFrag = null;      //定义Fragment变量
    private Fragment mContactFrag = null;
    private Fragment mFindFrag = null;
    private Fragment mSettingFrag = null;

    protected void onCreate(Bundle savedInstanceState) {
        super.onCreate(savedInstanceState);
        setContentView(R.layout.activity_main);
        initWidget();
    }
    private void initWidget()
```

```java
        {
            mFriendFrag = new FragmentFriend();        //创建一个FragmentFriend对象
            FragmentTransaction ft = getFragmentManager().beginTransaction();    //开启一个事务
            ft.replace(R.id.lly_content, mFriendFrag);        //初始显示聊天记录页面
            ft.commit();        //事务提交
            Button btnFriend = (Button)findViewById(R.id.btn_friend);        //获取按钮
            Button btnContact = (Button)findViewById(R.id.btn_contact);
            Button btnFind = (Button)findViewById(R.id.btn_find);
            Button btnSetting = (Button)findViewById(R.id.btn_setting);
            btnFriend.setOnClickListener(this);        //为按钮添加单击事件监听
            btnContact.setOnClickListener(this);
            btnFind.setOnClickListener(this);
            btnSetting.setOnClickListener(this);
        }
        @Override
        public void onClick(View view) {
            FragmentTransaction ft = getFragmentManager().beginTransaction();    //开启一个事务
            switch(view.getId())
            {
                case R.id.btn_friend:        //聊天记录
                {
                    if(null == mFriendFrag)
                    {
                        mFriendFrag = new FragmentFriend();        //创建对应的Fragment对象
                    }
                    ft.replace(R.id.lly_content, mFriendFrag);        //调用replace()方法替换当前Fragment
                    break;
                }
                case R.id.btn_contact:        //通讯录
                {
                    if(null == mContactFrag)
                    {
                        mContactFrag = new FragmentContact();
                    }
                    ft.replace(R.id.lly_content, mContactFrag);
                    break;
                }
```

```
            case R.id.btn_find:        //发现
            {
                if(null == mFindFrag)
                {
                    mFindFrag = new FragmentFind();
                }
                ft.replace(R.id.lly_content, mFindFrag);
                break;
            }
            case R.id.btn_setting:      //设置
            {
                if(null == mSettingFrag)
                {
                    mSettingFrag = new FragmentSetting();
                }
                ft.replace(R.id.lly_content, mSettingFrag);
                break;
            }
            default:
                break;
        }
        ft.commit();
    }
}
```

activity_main.xml 代码：

```
<LinearLayout xmlns:android="http://schemas.android.com/apk/res/android"
    xmlns:tools="http://schemas.android.com/tools"
    android:layout_width="match_parent"
    android:layout_height="match_parent"
    android:gravity="center"
    android:orientation="vertical" >
    <FrameLayout       //定义一个组件占用该位置,用于显示Fragment
        android:id="@+id/lly_content"
        android:layout_width="match_parent"
        android:layout_height="400dp"
        />
    <LinearLayout
```

```
        android:layout_width="match_parent"
        android:layout_height="50dp"
        android:orientation="horizontal"
        android:gravity="center">
        <Button      //聊天记录按钮
            android:id="@+id/btn_friend"
            android:layout_width="0dp"
            android:layout_height="wrap_content"
            android:layout_weight="1"
            android:textSize="16sp"
            android:text="friend"/>
        <Button      //通讯录按钮
            android:id="@+id/btn_contact"
            android:layout_width="0dp"
            android:layout_height="wrap_content"
            android:layout_weight="1"
            android:textSize="16sp"
            android:text="contact"/>
        <Button      //发现按钮
            android:id="@+id/btn_find"
            android:layout_width="0dp"
            android:layout_height="wrap_content"
            android:layout_weight="1"
            android:textSize="16sp"
            android:text="find"/>
        <Button      //设置（我）按钮
            android:id="@+id/btn_setting"
            android:layout_width="0dp"
            android:layout_height="wrap_content"
            android:layout_weight="1"
            android:textSize="16sp"
            android:text="setting"/>
    </LinearLayout>
</LinearLayout>
```

- **案例分析**

本例为每一种内容都定义了一个 Fragment，在 Fragment 的 onCreateView()方法中可以加载自己想要显示的布局文件，布局的处理并无特殊。然后在 Activity 中可以动态地控制 Fragment 的显示，非常灵活。

8.5.2 Fragment 的生命周期

Fragment 有自己的生命周期，但是其生命周期又依赖于 Activity。当 Activity 处于不可见状态时，Fragment 也一定处于不可见状态。当 Activity 处于销毁状态时，Fragment 也一定处于销毁状态。Fragment 在各个生命周期阶段也有不同的回调方法，具体流程如图 8.5 所示。

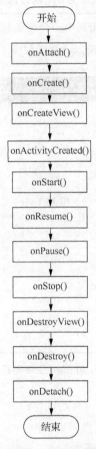

图 8.5　Fragment 的生命周期回调方法

从图 8.5 中可以看出，Fragment 的生命周期方法有些和 Activity 一致，意义也一致。表 8.11 用于说明 Fragment 特有的生命周期方法的含义。

表 8.11　Fragment 的生命周期方法说明

方法名	方法说明
onAttach()	Fragment 与 Activity 关联时调用该方法
onCreateView()	创建 Fragment 的视图，开发者在其中加载布局

续表

方法名	方法说明
onActivityCreated()	Activity 的 onCreate()方法已经返回时回调
onDestroyView()	销毁 Fragment 的视图时回调
onDetach()	取消 Fragment 与 Activity 的关联时回调

下面我们通过日志打印具体看一下 Fragment 生命周期方法的回调顺序。

案例 8.8　通过日志打印看 Fragment 生命周期方法的回调顺序

本例在 Activity 中显示两个按钮，单击其中一个按钮，将一个 Fragment 添加到 Activity 中，单击另外一个按钮，将 Fragment 从 Activity 中移除，界面如图 8.6 所示。

图 8.6　Fragment 生命周期实例界面

- **案例代码**

MainActivity 代码：

```
public class MainActivity extends Activity implements OnClickListener{
    private MyFragment mFragment = null;
    @Override
    protected void onCreate(Bundle savedInstanceState) {
        super.onCreate(savedInstanceState);
        setContentView(R.layout.activity_main);
        initWidget();
    }
    private void initWidget()
```

```java
        {
            Button btnAdd = (Button)findViewById(R.id.btn_add);        //获取按钮
            Button btnRemove = (Button)findViewById(R.id.btn_remove);
            btnAdd.setOnClickListener(this);        //添加单击监听事件
            btnRemove.setOnClickListener(this);
        }
        @Override
        public void onClick(View v) {
            FragmentTransaction ft = getFragmentManager().beginTransaction();    //开启一个事务
            switch(v.getId())
            {
            case R.id.btn_add: //添加Fragment
            {
                if(null == mFragment)
                {
                    mFragment = new MyFragment();        //定义一个Fragment对象
                }
                ft.add(R.id.fragment, mFragment);        //将Fragment添加到指定位置
                break;
            }
            case R.id.btn_remove:        //移除Fragment
            {
                ft.remove(mFragment);
                break;
            }
            default:
                break;
            }
            ft.commit();
        }
    }
```

MyFragment 代码：

```java
public class MyFragment extends Fragment {
    @Override
    public void onActivityCreated(Bundle savedInstanceState) {
        super.onActivityCreated(savedInstanceState);
        Log.i("Fragment", "onActivityCreated()");
```

```java
    }
    @Override
    public void onAttach(Activity activity) {
        super.onAttach(activity);
        Log.i("Fragment", "onAttach()");
    }
    @Override
    public void onCreate(Bundle savedInstanceState) {
        super.onCreate(savedInstanceState);
        Log.i("Fragment", "onCreate()");
    }
    @Override
    public View onCreateView(LayoutInflater inflater, ViewGroup container,
            Bundle savedInstanceState) {
        Log.i("Fragment", "onCreateView()");
        return inflater.inflate(R.layout.vew_fragment, container, false);
    }
    public void onDestroy() {
        Log.i("Fragment", "onDestroy()");
        super.onDestroy();
    }
    @Override
    public void onDestroyView() {
        Log.i("Fragment", "onDestroyView()");
        super.onDestroyView();
    }
    @Override
    public void onDetach() {
        Log.i("Fragment", "onDetach()");
        super.onDetach();
    }
    @Override
    public void onPause() {
        Log.i("Fragment", "onPause()");
        super.onPause();
    }
```

```java
    @Override
    public void onResume() {
        super.onResume();
        Log.i("Fragment", "onResume()");
    }
    @Override
    public void onStart() {
        super.onStart();
        Log.i("Fragment", "onStart()");
    }
    @Override
    public void onStop() {
        Log.i("Fragment", "onStop()");
        super.onStop();
    }
}
```

- **案例分析**

单击 Activity 中的 "add" 按钮，将 Fragment 添加到 Activity 中，控制台打印如图 8.7 所示。依次回调了 onAttach()、onCreate()、onCreateView()、onActivityCreated()、onStart()、onResume()方法，Fragment 处于前台可见。

按 "Home" 键，使 Activity 处于后台不可见状态，Fragment 也随之处于不可见状态，控制台打印如图 8.8 所示。依次回调了 onPause()和 onStop()方法。

单击 Activity 中的 "remove" 按钮，将 Fragment 移除，控制台打印如图 8.9 所示，依次回调了 onPause()、onStop()、onDestroyView()、onDestroy()、onDetach()方法。

L...	Time	PID	TID	Application	Tag	Text
I	07-04 06:42:09.269	666	666	com.demo.fragment	Fragment	onAttach()
I	07-04 06:42:09.269	666	666	com.demo.fragment	Fragment	onCreate()
I	07-04 06:42:09.269	666	666	com.demo.fragment	Fragment	onCreateView()
I	07-04 06:42:09.309	666	666	com.demo.fragment	Fragment	onActivityCreated()
I	07-04 06:42:09.329	666	666	com.demo.fragment	Fragment	onStart()
I	07-04 06:42:09.329	666	666	com.demo.fragment	Fragment	onResume()

图 8.7 添加 Fragment

I	07-04 06:43:13.579	666	666	com.demo.fragment	Fragment	onPause()
I	07-04 06:43:16.509	666	666	com.demo.fragment	Fragment	onStop()

图 8.8 按 "Home" 键

I	07-04 06:44:11.819	666	666	com.demo.fragment	Fragment	onPause()	
I	07-04 06:44:11.819	666	666	com.demo.fragment	Fragment	onStop()	
I	07-04 06:44:11.829	666	666	com.demo.fragment	Fragment	onDestroyView()	
I	07-04 06:44:11.829	666	666	com.demo.fragment	Fragment	onDestroy()	
I	07-04 06:44:11.839	666	666	com.demo.fragment	Fragment	onDetach()	

图 8.9　Fragment 销毁

8.6　RecyclerView

RecyclerView

在 Android 应用中，很多时候需要显示大量的数据，例如音乐播放器的歌曲列表，手机中可能存在大量的歌曲文件，但是同一时刻屏幕上只显示部分歌曲，用户可以通过上下滑动查看更多音乐。这种应用场景可以使用 2.2 节介绍的 ListView 和 GridView 实现布局。但是，Android 还提供了一种更灵活的组件——RecyclerView，它可以动态地实现列表布局或网格布局，甚至每个 Item 都可以显示不同的布局文件。而且，RecyclerView 内部还实现了 View 的复用，在用户上下滑动列表或网格时，RecyclerView 并不会为每一个子项创建一个 View，而是创建若干个 View 不断复用，更新其中显示的数据。

8.6.1　RecyclerView 相关类

在 RecyclerView 视图中，每个子项都被表示为 ViewHolder 对象，开发者必须继承 RecyclerView.ViewHolder 类实现自定义的 ViewHolder 类。每个 ViewHolder 都可以有一个布局文件用于显示一个子项。例如，如果用 RecyclerView 显示音乐列表，则每一个 ViewHolder 可以代表一首歌曲，用来显示歌曲的歌手信息、时长信息等，还可以响应单击和长按事件等。

RecyclerView 通过 Adapter 管理 ViewHolder，开发者必须继承 RecyclerView.Adapter 类实现自定义的 Adapter。Adapter 中有很多回调方法，其中 onCreateViewHolder()方法用于创建一个对应的 ViewHolder，onBindViewHolder()用于将数据和 ViewHolder 进行绑定。

RecyclerView 使用 LayoutManager 对其中的数据进行排列，开发者可以使用系统提供的 LayoutManager，也可以自定义 LayoutManager 类实现自己想要的排列方式。系统提供的 LayoutManager 有 3 种，分别是 LinearLayoutManager、GridLayoutMananger 和 StaggeredGridLayoutManager。LinearLayoutManager 将 RecyclerView 中的子项以一维列表的形式排列，类似于 ListView 的显示方式。GridLayoutManager 将 RecyclerView 中的子项以二维网格的形式排列，类似于 GridView 的显示方式。StaggeredGridLayoutManager 以瀑布流的方式排列 RecyclerView 中的子项。

8.6.2　RecyclerView 的使用

介绍完 RecyclerView 的一些基本概念，下面我们通过一个实例说明一下 RecyclerView 的具体使用，读者也可以和第 2 章介绍的 ListView 和 GridView 做个对比，体会 RecyclerView 的灵活性。

第 8 章 高级编程

案例 8.9 使用 RecyclerView

下面我们通过一个实例说明一下 RecyclerView 的具体使用，运行结果如图 8.10 所示。

图 8.10 RecyclerView 的使用

- **案例代码**

MainActivity 代码：

```java
public class MainActivity extends Activity {
    private RecyclerView mRecyclerView;      //定义一个RecyclerView对象
    private CustomAdapter mAdapter;           //定义一个Adapter对象
    @Override
    protected void onCreate(Bundle savedInstanceState) {
        super.onCreate(savedInstanceState);
        setContentView(R.layout.activity_main);
        initWidget();
        initData();
    }
    private void initWidget()
    {
        mRecyclerView = (RecyclerView)findViewById(R.id.recyclerview);     //获取RecyclerView
        LinearLayoutManager layoutManager = new LinearLayoutManager(this, LinearLayoutManager.VERTICAL, false);     //定义一个LayoutManager对象
        mRecyclerView.setLayoutManager(layoutManager);
        mAdapter = new CustomAdapter(this);      //新建一个Adapter对象
```

```java
        mRecyclerView.setAdapter(mAdapter);
    }
    private void initData()      //构造数据用于显示
    {
        List<Music> musics = new ArrayList<Music>();
        Music music1 = new Music("青花瓷", "周杰伦", "04:30");
        Music music2 = new Music("海阔天空", "Beyond", "05:25");
        Music music3 = new Music("下雨天", "南拳妈妈", "04:13");
        musics.add(music1);
        musics.add(music2);
        musics.add(music3);
        mAdapter.updateData(musics);
    }
}
```

activity_main.xml 代码：

```xml
<RelativeLayout xmlns:android="http://schemas.android.com/apk/res/android"
    xmlns:app="http://schemas.android.com/apk/res-auto"
    xmlns:tools="http://schemas.android.com/tools"
    android:layout_width="match_parent"
    android:layout_height="match_parent"
    >
    <android.support.v7.widget.RecyclerView   //定义一个RecyclerView布局
        android:id="@+id/recyclerview"
        android:layout_width="match_parent"
        android:layout_height="match_parent"
        android:layout_marginLeft="10dp"
        android:layout_marginRight="10dp" />
</RelativeLayout>
```

CustomAdapter 代码：

```java
//定义一个CustomAdapter类的对象，继承自RecyclerView.Adapter类
public class CustomAdapter extends RecyclerView.Adapter<RecyclerView.ViewHolder> {
    private Context mContext;
    private List<Music> mData = new ArrayList<Music>();
    public CustomAdapter(Context context)
    {
        mContext = context;
    }
```

```java
public void updateData(List<Music> data)      //对外提供方法用于刷新数据
{
    mData.addAll(data);
    notifyDataSetChanged();
}
public int getItemCount() {
    if(null == mData)
    {
        return 0;
    }
    return mData.size();
}
//重写onBindViewHolder方法,将数据和ViewHolder绑定
public void onBindViewHolder(ViewHolder holder, int position) {
    if(holder instanceof CustomHolder)
    {
        Music music = mData.get(position);
        ((CustomHolder) holder).txtTitle.setText(music.getTitle());
        ((CustomHolder) holder).txtActor.setText(music.getActor());
        ((CustomHolder) holder).txtTime.setText(music.getActor());
    }
}
//重写onCreateViewHolder方法,返回一个ViewHolder对象
public ViewHolder onCreateViewHolder(ViewGroup parent, int position) {
    View v = LayoutInflater.from(mContext).inflate(R.layout.vew_item, parent, false);
    return (new CustomHolder(v));
}
class CustomHolder extends RecyclerView.ViewHolder{  //自定义ViewHolder
    private TextView txtTitle;      //定义控件用于显示数据
    private TextView txtActor;
    private TextView txtTime;
    public CustomHolder(View v)
    {
        super(v);
        txtTitle = (TextView)v.findViewById(R.id.txt_title);
        txtActor = (TextView)v.findViewById(R.id.txt_actor);
        txtTime = (TextView)v.findViewById(R.id.txt_time);
```

```
        }
    }
}
```

- 案例分析

从实例中可以看到，RecyclerView 在配置文件中的使用与普通组件一样，然后在代码中调用了 findViewById()方法获取 RecyclerView 控件。在自定义的 CustomAdapter 类中，重写了父类 RecyclerView.Adapter 中的 onCreateViewHolder()和 onBindViewHolder()方法。在 onCreateViewHolder 中，创建并返回了一个自定义的 ViewHolder 对象，在 onBindViewHolder()方法中，将数据和 ViewHolder 绑定并显示。

8.7 本章小结

本章我们主要扩展地介绍了一些进阶知识，随着移动互联网的高速发展，移动应用几乎都需要一些网络通信。本章我们首先介绍了 Android 中的网络编程，然后介绍了一些图形图像和动画的知识，可以丰富 Android 页面的显示。8.3 节介绍了 MediaPlayer 类，可以用于播放一些音/视频。在 Android 应用中，经常会有一些比较耗时的任务，本章我们介绍了 AsyncTask 用于异步刷新，ThreadPoolExecutor 用于管理线程。最后介绍的 Fragment 可以支持更加灵活的碎片式开发，RecyclerView 可以方便用户更好地自定义实现 ListView 和 GridView。在掌握基础知识的同时，读者多了解一些新的开发技术和控件，可以更灵活地开发出更丰富的界面和功能。

第9章

综合实战

■ 在前面几章中我们介绍了 Android 开发的一些基础知识，在本章我们将综合运用之前的一些知识实现一个常见的视频播放器，使其可以播放本地的视频资源。

9.1 视频播放器

综合实例

对于一个视频播放器来说，常见的功能有进度条显示和拖动功能，用户拖动进度条可以快进和快退，在播放的过程中，进度条应该不断地更新播放的时间和显示视频总时长，同时还需要提供暂停/继续播放功能。其播放和显示功能主要通过第 8 章介绍的 MediaPlayer 和 SurfaceView 来实现，下面我们将分别描述不同的模块。

9.1.1 界面布局

界面主体部分需要添加一个 SurfaceView 用于显示播放的内容，然后需要显示进度条和播控按钮。为了可以更灵活地布局这些组件，布局方式可以采用 RelativeLayout，具体代码如下：

```xml
<RelativeLayout xmlns:android="http://schemas.android.com/apk/res/android"
    xmlns:tools="http://schemas.android.com/tools"
    android:layout_width="match_parent"
    android:layout_height="wrap_content"
    android:layout_gravity="center_horizontal" >
    <SurfaceView
        android:id="@+id/svew"
        android:layout_width="match_parent"
        android:layout_height="match_parent" />

    <LinearLayout
        android:id="@+id/llview"
        android:layout_width="match_parent"
        android:layout_height="wrap_content"
        android:layout_alignBottom="@id/svew"
        android:background="#777777"
        android:orientation="vertical"
        android:paddingBottom="15dp"
        android:alpha="0.8">

        <SeekBar
            android:id="@+id/seek_bar"
            android:layout_width="match_parent"
            android:layout_height="wrap_content"
            android:indeterminate="false"/>
```

```xml
<LinearLayout
    android:id="@+id/llview_progress"
    android:layout_width="wrap_content"
    android:layout_height="wrap_content"
    android:layout_marginTop="5dp"
    android:layout_marginBottom="5dp"
    android:layout_marginLeft="5dp"
    android:gravity="left"
    android:orientation="horizontal">

    <ImageView
        android:id="@+id/img_play_pause"
        android:layout_width="30dp"
        android:layout_height="30dp"
        android:clickable="true"
        android:src="@drawable/pause"/>

    <TextView
        android:id="@+id/txt_cur_time"
        android:layout_width="wrap_content"
        android:layout_height="wrap_content"
        android:layout_marginLeft="10dp"
        android:text="00:00:00"
        android:textColor="#FFFFFF"
        android:textSize="18sp" />
    <TextView
        android:id="@+id/txt_total_time"
        android:layout_width="wrap_content"
        android:layout_height="wrap_content"
        android:text="00:00:00"
        android:textColor="#FFDDBB"
        android:textSize="18sp"/>
    </LinearLayout>
    </LinearLayout>
</RelativeLayout>
```

在布局文件中，外层的视图容器为相对布局方式，其中添加了一个 SurfaceView 视图，layout_width

和 layout_height 的值都为 match_parent，使得播放内容占满整个画面。其后定义了一个垂直的线性布局，用于显示进度条和播控按钮等内容，通过设置 android:layout_alignBottom="@id/svew"使得该布局与 SurfaceView 的底部对齐，设置 android:alpha 属性让其具有一定的透明度。在该布局中，首先定义了一个 SeekBar 用于显示播放的进度和拖动条，然后定义了一个水平的线性布局用于显示播控按钮和播放时间。播控按钮通过设置 android:clickable="true"使得该图片可以被单击，用于实现播放/暂停功能。页面具体效果如图 9.1 所示。

图 9.1 页面布局

9.1.2 初始化

视频播放需要指定视频所在的路径或 URL，本章我们基于第 5 章小练习的文件浏览器获取视频路径，在文件浏览的过程中，如果是文件夹，则进入下级目录，如果是视频文件，则调用本章的视频播放器，并将视频的路径作为参数传入，其他文件则返回。代码如下所示：

```
private void change(File file)
{
    String fileName = file.getName();
    if(fileName.endsWith(".avi") || fileName.endsWith(".mp4") || fileName.endsWith(".mkv"))
    {
        // 调用本章的视频播放页面
        Intent intent = new Intent(MainActivity.this, VideoPlayerActivity.class);
        intent.putExtra("videoUrl", file.getAbsolutePath());
        startActivity(intent);
    }
    if(!file.isDirectory())
    {
        return;
```

```
        }
        mTitle.setText(file.getAbsolutePath());
        List<File> files = fileMgr.getSubFiles(file);
        mAdpter.updateFiles(files);
        mAdpter.notifyDataSetChanged();
    }
```

视频播放 Activity 启动时，首先在 onCreate()方法中设置一些 Windows 参数使得播放页面全屏显示，然后从 intent 中获取视频的 URL，并做一些组件和播放器的初始化工作，具体如下：

```
    @Override
    protected void onCreate(Bundle savedInstanceState) {
        super.onCreate(savedInstanceState);
        this.requestWindowFeature(Window.FEATURE_NO_TITLE);
        this.getWindow().setFlags(WindowManager.LayoutParams.FLAG_FULLSCREEN,
                    WindowManager.LayoutParams.FLAG_FULLSCREEN);
        setContentView(R.layout.activity_video);
        //获取视频的路径
        Intent intent = getIntent();
        mVideoUrl = intent.getStringExtra("videoUrl");

        initWidget();
        initPlayer();
        mHandler.postDelayed(task, 1000);
    }

    /**
     * 初始化控件
     **/
    private void initWidget()
    {
        mSurface = (SurfaceView)findViewById(R.id.svew);
        mSeekBar = (SeekBar)findViewById(R.id.seek_bar);
        mtxtCurTime = (TextView)findViewById(R.id.txt_cur_time);
        mtxtTotalTime = (TextView)findViewById(R.id.txt_total_time);
        mPlayPause = (ImageView)findViewById(R.id.img_play_pause);

        //设置播放进度的初始值
        mtxtCurTime.setText("00:00:00" + " / ");
```

```java
            mtxtTotalTime.setText("00:00:00");
            mSeekBar.setProgress(0);
            mSeekBar.setMax(100);
            //播控按钮增加单击事件监听
            mPlayPause.setOnClickListener(new OnClickListener(){
                @Override
                public void onClick(View v) {
                    if(mPlayer.isPlaying())
                    {
                        mPlayer.pause();
                        mPlayPause.setImageResource(R.drawable.play);
                    }
                    else
                    {
                        mPlayer.start();
                        mPlayPause.setImageResource(R.drawable.pause);
                    }
                }});

            mSeekBar.setOnSeekBarChangeListener(this);
    }

    /**
     * 初始化播放器
     **/
    private void initPlayer()
    {
        mPlayer = new MediaPlayer();
        try{
            //设置视频源
            mPlayer.setDataSource(mVideoUrl);
            mHolder = mSurface.getHolder();
            // holder增加回调函数
            mHolder.addCallback(new SurfaceHolder.Callback() {
                @Override
                public void surfaceDestroyed(SurfaceHolder holder) {
```

```java
            }
            @Override
            public void surfaceCreated(SurfaceHolder holder) {
                mPlayer.setDisplay(holder);
            }
            @Override
            public void surfaceChanged(SurfaceHolder holder, int format, int width,
                    int height) {
            }
        });
        //播放器准备中
        mPlayer.prepareAsync();
        //设置播放器准备的回调函数
        mPlayer.setOnPreparedListener(new OnPreparedListener(){
            @Override
            public void onPrepared(MediaPlayer mp) {
                if(null != mp)
                {
                    mp.start(); //准备完成就播放
                    mVideoLength = mp.getDuration(); //获取视频总时长并显示
                    mtxtTotalTime.setText(CommonUtils.formatVideoLegth(mVideoLength));
                }
            }
        });
    }catch(Exception e)
    {
        e.printStackTrace();
    }
}
```

在组件的初始化函数 initWidget()方法中，首先通过 findViewById()方法获取各个控件，然后为进度的显示和进度条的位置设置初始值。通过 setOnClickListener 方法为播控图片设置监听事件。

在播放器的初始化函数 initPlayer()方法中，首先创建了一个 MediaPlayer 类，通过 setDataSource() 方法设置视频源，然后获取 SurfaceView 的 SurfaceHolder，为其添加回调函数。最后调用 MediaPlayer 类的 prepareAsync()方法异步准备播放源，当播放准备完毕后，会回调 OnPreparedListener 的 onPrepared() 方法，在其中可以启动 MediaPlayer 对象进行播放，并且可以获取视频的总时长。

9.1.3 播控和进度控制

在视频播放中，暂停、继续播放和进度拖动是常见的操作。播控是通过监听图片的单击事件实现的，如下所示：

```
// 播控按钮增加单击事件监听
mPlayPause.setOnClickListener(new OnClickListener(){
    @Override
    public void onClick(View v) {
        if(mPlayer.isPlaying())
        {
            mPlayer.pause();
            mPlayPause.setImageResource(R.drawable.play);
        }
        else
        {
            mPlayer.start();
            mPlayPause.setImageResource(R.drawable.pause);
        }
    }});
```

在图片按钮被单击后，首先判断当前视频的状态，如果是播放状态，则调用 MediaPlayer 类的 pause()方法暂停播放，同时更换播控的图片，显示播放按钮。如果是暂停状态，则调用 MediaPlayer 类的 start()方法继续播放，同时更换播控的图片，显示暂停按钮。播放和暂停页面如图 9.2 和图 9.3 所示。

图 9.2　播放状态

图 9.3 暂停状态

拖动进度条进行快进和快退是通过监听 OnSeekBarChangeListener 事件实现的，首先在代码中通过 setOnSeekBarChangeListener() 为拖动条设置事件监听，然后重写 OnseekBarChangeListener 的回调函数，在不同的事件触发下执行不同的动作。代码具体如下：

```java
private void initWidget()
    {
        ……
        ……
        mSeekBar.setOnSeekBarChangeListener(this);
    }


/**
* 拖动条发生改变时回调
**/
@Override
public void onProgressChanged(SeekBar seekBar, int progress,
        boolean fromUser) {
    mPlayer.seekTo((int) ((progress / 100.0f) * mVideoLength));
}


 /**
* 触摸到拖动条时回调
**/
@Override
public void onStartTrackingTouch(SeekBar seekBar) {
```

```
            if(mPlayer.isPlaying())
            {
                mPlayer.pause();
            }
        }
        /**
         * 释放拖动条时回调
         **/
        @Override
        public void onStopTrackingTouch(SeekBar seekBar) {
            if(null != mPlayer && !mPlayer.isPlaying())
            {
                mPlayer.start();
            }
        }
```

在 OnseekBarChangeListener 的回调函数中，onProgressChanged 在拖动条发生改变时回调，并回调滑块当前的位置，在其中可以根据滑块的位置计算需要定位的视频位置，然后调用 MediaPlayer 的 seekTo()方法跳转到对应的位置。onStartTrackingTouch 在拖动条被触摸时回调，其中可以调用 MediaPlayer 的 pause()方法暂停视频的播放。onStopTrackingTouch 在拖动条被释放时回调，此时快进或快退已经完成，调用 MediaPlayer 的 start()方法继续播放。

在播放的过程中，需要根据当前视频播放的时间实时更新进度和时间显示的进度，可以通过 Handler 和 Runnable 任务实现，具体如下：

```
        private Runnable task = new Runnable()
        {
            @Override
            public void run() {
                int progress = mSeekBar.getProgress();
                int position = mPlayer.getCurrentPosition();
                if(0 != mVideoLength && mPlayer.isPlaying())
                {
                    //计算拖动条滑块应该处于的位置
                    progress = (int) ((position / (mVideoLength * 1.0f)) * 100);
                }
                mSeekBar.setProgress(progress);
                //更新当前应该显示的播放时间
                mtxtCurTime.setText(CommonUtils.formatVideoLegth(position) + " / ");
```

```
                if(progress >= 100)
                {
                    mHandler.removeCallbacks(this);
                }else
                {
                    mHandler.postDelayed(this, 1000);        //每隔1秒执行一次
                }
            }
        };
```

首先获取拖动条的当前进度，然后调用 MediaPlayer 的 getCurrentPosition()获取当前视频播放的位置，通过和总时长的比较计算拖动条应该处于的位置，调用 SeekBar 的 setProgress()的方法更新滑块的显示位置。另外将视频的当前的播放时长转化为 00:00:00 格式更新显示到文本框中。格式化时间的代码如下：

```
public static String formatVideoLegth(int videoLength)
{
    int mills = videoLength / 1000;          //毫秒转换成秒
    int seconds = mills % 60;                //获取秒
    int minutes = mills / 60 % 60;           //获取分钟
    int hours = mills / 3600;                //获取小时
    StringBuilder sBuilder = new StringBuilder();
    sBuilder.append(String.format("%02d", hours)).append(":");
    sBuilder.append(String.format("%02d", minutes)).append(":");
    sBuilder.append(String.format("%02d", seconds));           //格式化
    return sBuilder.toString();
}
```

9.1.4 横屏设置

在视频播放中，一般会自动切换到横屏显示，即视频的宽比高大，这可以在声明 Activity 时通过指定 android:screenOrientation 属性实现，具体如下：

```
<activity
    android:name="com.demo.fileexplorer.VideoPlayerActivity"
    android:screenOrientation="landscape">
    <intent-filter>
        <action android:name="android.intent.action.videoplayer" />
    </intent-filter>
</activity>
```

android:screenOrientation 属性支持 3 种取值，"landscape"强制横屏显示，"portrait"强制竖屏显示，默认值为"unspecified"，显示方向跟随系统屏幕旋转的方向。

9.2 本章小结

　　本章我们分模块介绍了一个视频播放页面的实现，首先介绍了界面的布局方式，可以看到相对布局和线性布局在实际开发中的灵活组合和应用。然后介绍了控件和播放器的初始化，回顾了通过 Intent 进行数据传递的操作。后面通过对播控和进度控制的分析，展示了 Android 中基于回调的事件监听机制和使用 Handler 实现在线程中异步刷新 UI 页面的操作。最后通过将 Activity 设置成横屏显示，熟悉了在 AndroidMainfest.xml 文件中对 Activity 进行属性设置的方法。需要指出的是，在 Android 的 xml 配置文件中，可以对界面的视图和各个组件做各种属性的设置，本书并不能一一涵盖，需要各位读者在开发过程中不断进行积累和熟悉。